The
Intercept_

ALSO BY JEREMY SCAHILL

Dirty Wars: The World Is a Battlefield
Blackwater: The Rise of the World's Most Powerful Mercenary Army

THE ASSASSINATION COMPLEX
INSIDE THE GOVERNMENT'S SECRET DRONE WARFARE PROGRAM

JEREMY SCAHILL AND THE STAFF OF THE INTERCEPT

Simon & Schuster Paperbacks
New York London Toronto Sydney New Delhi

Simon & Schuster Paperbacks
An Imprint of Simon & Schuster, Inc.
1230 Avenue of the Americas
New York, NY 10020

First Simon & Schuster trade paperback edition May 2017

Interior design: Philipp Hubert
Typeface: TI Actu by Stéphane Elbaz and Swift Neue by Gerard Unger

Manufactured in the United States of America

10 9 8 7 6 5 4 3 2

The Library of Congress has cataloged the hardcover edition as follows:

Names: Scahill, Jeremy, author.
Title: The assassination complex : inside the government's secret drone warfare program / by Jeremy Scahill and the staff of The Intercept.
Other titles: Intercept.
Description: New York : Simon & Schuster, [2016]
Identifiers: LCCN 2016001115| ISBN 9781501144134 | ISBN 1501144138
Subjects: LCSH: Drone aircraft – Government policy – United States. | Targeted Killing – United States – History – 21st century. | Targeted killing – Moral and ethical aspects – United States. | Drone aircraft – Moral and ethical aspects – United States. | Terrorism – Prevention – United States – Decision making. | Military intelligence – United States – Evaluation. | Military intelligence – Technological innovation – Moral and ethical aspects. | United States – Military policy – Moral and ethical aspects. | National Security – United States.
Classification: LCC UG1242.D7 S33 2016 | DDC 358.4 – dc23 LC record available at http://lccn.loc.gov/2016001115

ISBN 978-1-5011-4413-4
ISBN 978-1-5011-4414-1 (pbk)
ISBN 978-1-5011-4415-8 (ebook)

For those who speak out clearly
and pay up personally.

CONTENTS

PREFACE
BETSY REED

The story of "The Drone Papers," on which this book is based, began, as national security exposés often do, with a reporter and a source. Jeremy Scahill, whose previous books penetrated the secret world of government security contractors and special ops, was approached by a whistleblower from within the U.S. intelligence community whose conscience demanded that he reveal to the public the true nature of the covert war being waged in the name of their security.

What followed was a painstaking process of reporting and analysis by a team of reporters, researchers, editors, and designers at *The Intercept*, the digital investigative magazine that Scahill founded with Glenn Greenwald and Laura Poitras in the wake of NSA whistleblower Edward Snowden's massive leak of documents exposing the extent of domestic surveillance.

On October 15, 2015, *The Intercept* published "The Drone Papers," a comprehensive investigation based on a new set of documents, detailing the U.S. military's drone wars in Yemen, Somalia, and Afghanistan. Providing an unprecedented look at the military's drone-based assassination program, the series revealed that drone strikes kill far greater numbers of people than those on kill lists—in one Afghan campaign 90 percent of those killed were not the intended target—and that the military classifies unknown persons, often innocent bystanders, as "enemies killed in action."

In meticulously reported stories accompanied by vivid data visualizations, "The Drone Papers" illuminated the culture that celebrates targeted kills with terms like "jackpot" and "touchdown." Exposing for the first time an official "kill chain" leading all the way to the president, *The Intercept*'s reporting paints a careful yet devastating portrait of how the drone campaign harms U.S. intelligence and exacerbates the very threat the war on terror seeks to overcome.

"The Drone Papers" earned extensive broadcast coverage and was the focus of hundreds of follow-up stories worldwide. Most major newspapers covered the story, many declaring the emergence of a "new Edward Snowden." Published the same day President Obama declared that U.S. troops would remain in Afghanistan, "The Drone Papers" framed the reception of that announcement and sparked debate about America's role in the world.

The Assassination Complex presents the original articles from "The Drone Papers" along with additional reporting from *The Intercept* on the process by which the U.S. government creates its watchlists, the role of the National Security Agency in the assassination program, and the inevitable use of military surveillance technology in domestic policing. This body of reporting provides an unparalleled glimpse into the shadowy world of extrajudicial assassination that promises to be Barack Obama's most troubling legacy.

FOREWORD: ELECTED BY CIRCUMSTANCE
EDWARD SNOWDEN

"I've been waiting forty years for someone like you." Those were the first words Daniel Ellsberg spoke to me when we met last year. Dan and I felt an immediate kinship; we both knew what it meant to risk so much – and to be irrevocably changed – by revealing secret truths.

One of the challenges of being a whistleblower is living with the knowledge that people continue to sit, just as you did, at those desks, in that unit, throughout the agency, who see what you saw and comply in silence, without resistance or complaint. They learn to live not just with untruths but with *unnecessary* untruths, *dangerous* untruths, *corrosive* untruths. It is a double tragedy: what begins as a survival strategy ends with the compromise of the human being it sought to preserve and the diminishing of the democracy meant to justify the sacrifice.

But unlike Dan Ellsberg, I didn't have to wait forty years to witness other citizens breaking that silence with documents. Ellsberg gave the Pentagon Papers to the *New York Times* and other newspapers in 1971; Chelsea Manning provided the Iraq and Afghan War logs and the Cablegate materials to WikiLeaks in 2010. I came forward in 2013. Now here we are in 2015, and another person of courage and conscience has made available the set of extraordinary documents that are published here.

We are witnessing a compression of the working period in which bad policy shelters in the shadows, the time frame in which unconstitutional activities can continue before they are exposed by acts of conscience. And this temporal compression has a significance beyond the immediate headlines; it permits the people of this country to learn about critical government actions, not as part of the historical record but in a way that allows direct action through voting – in other words, in a way that empowers an informed citizenry to defend the democracy that "state secrets" are nominally intended to support. When I see individuals who are able to bring information forward, it gives me hope that we won't always be required to curtail the illegal activities of our government as if it were a constant task, to uproot official lawbreaking as routinely as we mow the grass. (Interestingly enough, that is how some have begun to describe remote killing operations, as "cutting the grass.")

A single act of whistleblowing doesn't change the reality that there are significant portions of the government that operate below the waterline, beneath the visibility of the public. Those secret activities will continue, despite reforms. But those who perform these actions now have to live with the fear that if they engage in activities contrary to the spirit of society – if even a single citizen is catalyzed to halt the machinery of that injustice – they might still be held to account. The thread by which good governance hangs is this equality before the law, for the only fear of the man who turns the gears is that he may find himself upon them.

Hope lies beyond, when we move from extraordinary acts of revelation to a collective culture of accountability within the intelligence community. Here we will have taken a meaningful step toward solving a problem that has existed for as long as our government.

Not all leaks are alike, nor are their makers. Gen. David Petraeus, for instance, provided his illicit lover and favorable biographer information so secret it defied classification, including the names of covert operatives and the president's private thoughts on matters of strategic concern. Petraeus was not charged with a felony, as the Justice Department had initially recommended, but was instead permitted to plead guilty to a misdemeanor. Had an enlisted soldier of modest rank pulled out a stack of highly classified notebooks and handed them to his girlfriend to secure so much as a smile, he'd be looking at many decades in prison, not a pile of character references from a Who's Who of the Deep State.

There are authorized leaks and also permitted disclosures. It is rare for senior administration officials to explicitly ask a subordinate to leak a CIA officer's name to retaliate against her husband, as appears to have been the case with Valerie Plame. It is equally rare for a month to go by in which some senior official does not disclose some protected information that is beneficial to the political efforts of the parties but clearly "damaging to national security" under the definitions of our law.

This dynamic can be seen quite clearly in the al Qaeda "conference call of doom" story, in which intelligence officials, likely seeking to inflate the threat of terrorism and deflect criticism of mass surveillance, revealed to a neoconservative website extraordinarily detailed accounts of specific communications they had intercepted, including locations of the participating parties and the precise contents of the discussions. If the officials' claims were to be believed, they irrevocably burned an extraordinary means of learning the precise plans and intentions of terrorist leadership for the sake of a short-lived political advantage in a news cycle. Not a single person seems to have been so much as disciplined as a result of the story that cost us the ability to listen to the alleged al Qaeda hotline.

If harmfulness and authorization make no difference, what explains the distinction between the permissible and the impermissible disclosure?

The answer is control. A leak is acceptable if it's not seen as a threat, as a challenge to the prerogatives of the institution. But if all of the disparate components of the institution – not just its head but its hands and feet, every part of its body – must be assumed to have the same power to discuss matters of concern, that is an existential threat to the modern political monopoly of information control, particularly if we're talking about disclosures of serious wrongdoing, fraudulent activity, unlawful activities. If you can't guarantee that you alone can exploit the flow of controlled information, then the aggregation of all the world's unmentionables – including your own – begins to look more like a liability than an asset.

Truly unauthorized disclosures are necessarily an act of resistance – that is, if they're not done simply for press consumption, to fluff up the public appearance or reputation of an institution. However, that doesn't mean they all come from the lowest working level. Sometimes the individuals who step forward happen to be near the pinnacle of power. Ellsberg was in the top tier; he was briefing the secretary of defense. You can't get much higher, unless you are the secretary of defense, and the incentives simply aren't there for such a high-ranking official to be involved in public interest disclosures because that person already wields the influence to change the policy directly.

At the other end of the spectrum is Manning, a junior enlisted soldier, who was much nearer to the bottom of the hierarchy. I was midway in the professional career path. I sat down at the table with the chief information officer of the CIA, and I was briefing him and his chief technology officer when they were publicly making statements like "We try to collect everything and hang on to it forever," and everybody still thought that was a cute business slogan. Meanwhile I was designing the systems they would use to do precisely that. I wasn't briefing the policy side, the secretary of defense, but I was briefing the operations side, the National Security Agency's director of technology. Official wrongdoing can catalyze all levels of insiders to reveal information, even at great risk to themselves, so long as they can be convinced that it is necessary to do so.

Reaching those individuals, helping them realize that their first allegiance as a public servant is to the public rather than to the government, is the challenge. That's a significant shift in cultural thinking for a government worker today.

I've argued that whistleblowers are elected by circumstance. It's not a virtue of who you are or your background. It's a question of what you are exposed to, what you witness. At that point the question becomes *Do you honestly believe that you have the capability to remediate the problem, to influence policy?* I would not encourage individuals to reveal information, even about wrongdoing, if they do not believe they can be effective in doing so, because the right moment can be as rare as the will to act.

This is simply a pragmatic, strategic consideration. Whistleblowers are outliers of probability, and if they are to be effective as a political force, it's critical that they maximize the amount of public good produced from scarce seed. When I was making my decision, I came to understand how one strategic consideration, such as waiting until the month before a domestic election, could become overwhelmed by another, such as the moral imperative to provide an opportunity to arrest a global trend that had already gone too far. I was focused on what I saw and on my sense of overwhelming disenfranchisement that the government, in which I had believed for my entire life, was engaged in such an extraordinary act of deception.

At the heart of this evolution is that whistleblowing is a radicalizing event – and by "radical" I don't mean "extreme"; I mean it in the traditional sense of *radix*, the root of the issue. At some point you recognize that you can't just move a few letters around on a page and hope for the best. You can't simply report this problem to your supervisor, as I tried to do, because inevitably supervisors get nervous. They think about the structural risk to their career. They're concerned about rocking the boat and "getting a reputation." The incentives aren't there to produce meaningful reform. Fundamentally, in an open society, change has to flow from the bottom to the top.

As someone who works in the intelligence community, you've given up a lot to do this work. You've happily committed yourself to tyrannical restrictions. You voluntarily undergo polygraphs; you tell the government everything about your life. You waive a lot of rights because you believe the fundamental goodness of your mission justifies the sacrifice of even the sacred. It's a just cause.

And when you're confronted with evidence – not in an edge case, not in a peculiarity, but as a core consequence of the program – that the government is subverting the Constitution and violating the ideals you so fervently believe in, you have to make a decision. When you see that the program or policy is inconsistent with the oaths and obligations that you've sworn to your society and yourself, then that oath and that obligation cannot be reconciled with the program. To which do you owe a greater loyalty?

One of the extraordinary things about the revelations of the past several years, and their accelerating pace, is that they have occurred in the context of the United States as the "uncontested hyperpower." We now have the largest unchallenged military machine in the history of the world, and it's backed by a political system that is increasingly willing to authorize any use of force in response to practically any justification. In today's context that justification is terrorism, but not necessarily because our leaders are particularly concerned about terrorism in itself or because they think it's an existential threat to society. They recognize that even if we had a 9/11 attack every year, we would still be losing more people to car accidents and heart disease, and we don't see the same expenditure of resources to respond to those more significant threats.

What it really comes down to is the political reality that we have a political class that feels it must inoculate itself against allegations of weakness. Our politicians are more fearful of the politics of terrorism – of the charge that they do not take terrorism seriously – than they are of the crime itself.

As a result we have arrived at this unmatched capability, unrestrained by policy. We have become reliant upon what was intended to be the limitation of last resort: the courts. Judges, realizing that their decisions are suddenly charged with much greater political importance and impact than was originally intended, have gone to great lengths in the post-9/11 period to avoid reviewing the laws or the operations of the executive in the national security context and setting restrictive precedents that, even if entirely proper, would impose limits on government for decades or more. That means the most powerful institution that humanity has ever witnessed has also become the least restrained. Yet that same institution was never designed to operate in such a manner, having instead been explicitly founded on the principle of checks and balances. Our founding impulse was to say, "Though we are mighty, we are voluntarily restrained."

When you first go on duty at CIA headquarters, you raise your hand and swear an oath – not to government, not to the agency, not to secrecy. You swear an oath to the Constitution. So there's this friction, this emerging contest between the obligations and values that the government asks you to uphold, and the actual activities that you're asked to participate in.

These disclosures about the Obama administration's killing program reveal that there's a part of the American character that is deeply concerned with the unrestrained, unchecked exercise of power. And there is no greater or clearer manifestation of unchecked power than assuming for oneself the authority to execute an individual outside of a battlefield context and without the involvement of any sort of judicial process.

Traditionally, in the context of military affairs, we've always understood that lethal force in battle could not be subjected to ex ante judicial constraints. When armies are shooting at each other, there's no room for a judge on that battlefield. But now the government has decided – without the public's participation, without our knowledge and consent – that the battlefield is everywhere. Individuals who don't represent an imminent threat in any meaningful sense of those words are redefined, through the subversion of language, to meet that definition.

Inevitably that conceptual subversion finds its way home, along with the technology that enables officials to promote comfortable illusions about surgical killing and nonintrusive surveillance. Take, for instance, the Holy Grail of drone persistence, a capability that the United States has been pursuing forever. The goal is to deploy solar-powered drones that can loiter in the air for weeks without coming down. Once you can do that, and you put any typical signals-collection device on the bottom of it to monitor, unblinkingly, the emanations of, for example, the different network addresses of every laptop, smartphone, and iPod, you know not just where a particular device is in what city, but you know what apartment each device lives in, where it goes at any particular time, and by what route. Once you know the devices, you know their owners. When you start doing this over several cities, you're tracking the movements not just of individuals but of whole populations.

By preying on the modern necessity to stay connected, governments can reduce our dignity to something like that of tagged animals, the primary difference being that we paid for the tags and they're in our pockets. It sounds like fantasist paranoia, but on the technical level it's so trivial to implement that I cannot imagine a future in which it won't be attempted. It will be limited to the war zones at first, in accordance with our customs, but surveillance technology has a tendency to follow us home.

Here we see the double edge of our uniquely American brand of nationalism. We are raised to be exceptionalists, to think we are the better nation with the manifest destiny to rule. The danger is that some people will actually believe this claim, and some of those will expect the manifestation of our national identity, that is, our government, to comport itself accordingly.

Unrestrained power may be many things, but it's not American. It is in this sense that the act of whistleblowing increasingly has become an act of political resistance. The whistleblower raises the alarm and lifts the lamp, inheriting the legacy of a line of Americans that begins with Paul Revere.

The individuals who make these disclosures feel so strongly about what they have seen that they're willing to risk their lives and their freedom. They know that we, the people, are ultimately the strongest and most reliable check on the power of government. The insiders at the highest levels of government have extraordinary capability, extraordinary resources, tremendous access to influence, and a monopoly on violence, but in the final calculus there is but one figure that matters: the individual citizen.

And there are more of us than there are of them.

THE ASSASSINATION COMPLEX

THE DRONE LEGACY

JEREMY SCAHILL

From his first days as commander in chief, the drone has been President Barack Obama's weapon of choice, used by the military and the CIA to hunt down and kill the people his administration has deemed—through secretive processes, without indictment or trial—deserving of execution. There has been intense focus on the technology of remote killing, but that often serves as a surrogate for what should be a broader examination of the state's power over life and death.

Drones are a tool, not a policy. The policy is assassination. While every president since Gerald Ford has upheld an executive order banning assassinations by U.S. personnel, Congress has avoided legislating the issue or even defining the word "assassination."[1] This has allowed proponents of the drone wars to rebrand assassinations with more palatable characterizations, such as the term du jour, "targeted killings."

When the Obama administration has discussed drone strikes publicly, it has offered assurances that such operations are a more precise alternative to boots on the ground and are authorized only when an "imminent" threat is present and there is "near certainty" that the intended target will be eliminated. Those terms, however, appear to have been bluntly redefined to bear almost no resemblance to their commonly understood meanings.[2]

The first drone strike outside of a declared war zone was conducted in 2002, yet it was not until May 2013 that the White House released a set of standards and procedures for conducting such strikes.[3] Those guidelines offered little specificity, asserting that the United States would conduct a lethal strike outside an "area of active hostilities" only if a target represents a "continuing, imminent threat to U.S. persons," without providing any sense of the internal process used to determine whether a suspect should be killed without being indicted or tried.[4] The implicit message on drone strikes from the Obama administration has been *Trust, but don't verify*.[5]

On October 15, 2015, *The Intercept* published a cache of secret slides that provide a window into the inner workings of the U.S. military's kill/capture operations during a key period in the evolution of the drone wars: between 2011 and 2013. The documents, which also outline the internal views of special operations forces on the shortcomings and flaws of the drone program, were provided by a source within the intelligence community who worked on the types of operations and programs described in the slides. We granted the source's request for anonymity because the materials are classified and because the U.S. government has engaged in aggressive prosecution of whistleblowers. Throughout this book, we will refer to this person simply as "the source."

The source said he decided to disclose these documents because he believes the public has a right to understand the process by which people are placed on kill lists and ultimately assassinated on orders from the highest echelons of the U.S. government: "This outrageous explosion of watchlisting, of monitoring people and racking and stacking them on lists, assigning them numbers, assigning them 'baseball cards,' assigning them death sentences without notice, on a worldwide battlefield, was, from the very first instance, wrong.

"We're allowing this to happen. And by 'we,' I mean every American citizen who has access to this information now, but continues to do nothing about it."

The Pentagon, White House, and Special Operations Command declined to comment on the documents. A Defense Department spokesperson said, "We don't comment on the details of classified reports."

The CIA and the U.S. military's Joint Special Operations Command (JSOC) operate parallel drone-based assassination programs, and the secret documents should be viewed in the context of an intense turf war over which entity should have supremacy in those operations. Two sets of slides focus on the military's high-value targeting campaign in Somalia and Yemen as it existed between 2011 and 2013, specifically the operations of a secretive unit, Task Force 48-4.[6] Additional documents on high-value kill/capture operations in Afghanistan buttress previous accounts of how the Obama administration masks the true number of civilians killed in drone strikes by categorizing unidentified people killed in a strike as enemies, even if they were not the intended targets.[7] The slides also paint a picture of a campaign in Afghanistan aimed at eliminating not only al Qaeda and Taliban operatives but also members of other local armed groups. One slide, marked "Top Secret," shows how the terror "watchlist" appears on the terminals of personnel conducting drone operations, linking unique codes associated with cell phone SIM cards and handsets to specific individuals in order to geolocate them.[8]

The costs to intelligence gathering when suspected terrorists are killed rather than captured are outlined in the slides pertaining to Yemen and Somalia, which are part of a 2013 study conducted by a Pentagon entity, the Intelligence, Surveillance, and Reconnaissance Task Force. The ISR study lamented the limitations of the drone program, arguing for more advanced drones and other surveillance aircraft and the expanded use of naval vessels to extend the reach of surveillance operations necessary for targeted strikes. It also contemplated the establishment of new "politically challenging" airfields and recommended capturing and interrogating more suspected terrorists rather than killing them in drone strikes.

The ISR Task Force at the time was under the control of Michael Vickers, the undersecretary of defense for intelligence. A fierce proponent of drone strikes and a legendary paramilitary figure, Vickers had long pushed for a significant increase in the military's use of special operations forces. Key lawmakers viewed the ISR Task Force as an advocate for more surveillance platforms, like drones.[9]

The ISR study also reveals new details about the case of a British citizen, Bilal el-Berjawi, who was stripped of his citizenship before being killed in a U.S. drone strike in 2012. British and American intelligence had Berjawi under surveillance for several years as he

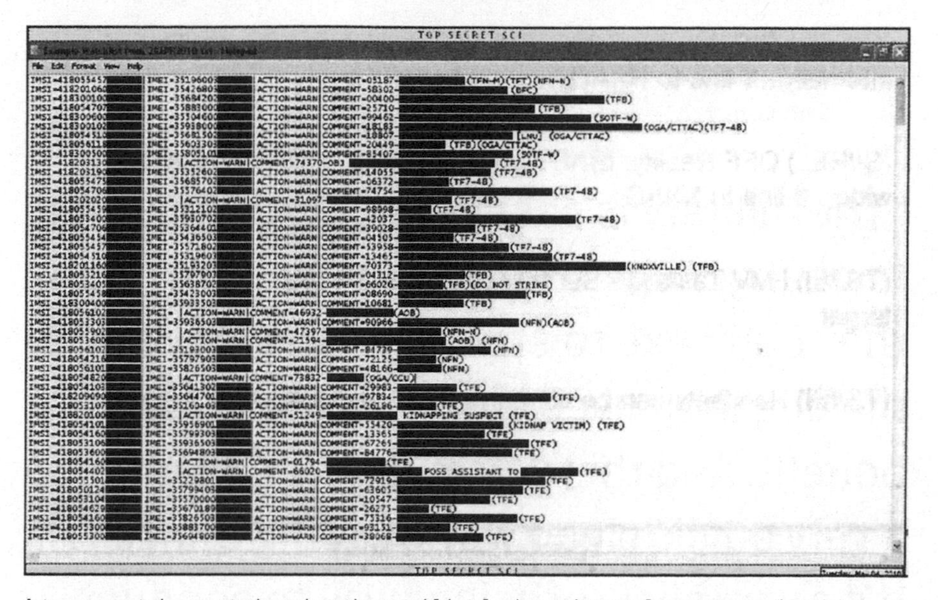

A top-secret document shows how the watchlist looks on internal systems used by drone operators.

traveled back and forth between the U.K. and East Africa yet did not capture him. Instead the United States hunted him down and killed him in Somalia.

Taken together, the secret documents lead to the conclusion that Washington's fourteen-year-long high-value targeting campaign suffers from an overreliance on signals intelligence, an apparently incalculable civilian toll, and, due to a preference for assassination rather than capture, an inability to extract potentially valuable intelligence from terror suspects. The documents also highlight the futility of the war in Afghanistan by showing that the United States has poured vast resources into killing local insurgents, in the process exacerbating the very threat the United States is seeking to confront.

These secret slides help provide a historical context to Washington's ongoing wars and are especially relevant today, as the U.S. military intensifies its drone strikes and covert actions against ISIS in Syria and Iraq.[10] Those campaigns, like the ones detailed in these documents, are unconventional wars that employ special operations forces at the tip of the spear. The "find, fix, finish" doctrine that has fueled a post-9/11 borderless war is being refined and institutionalized. These documents lay bare the normalization of assassination, whether through the use of drones, night raids, or new platforms yet to be unleashed, as a central component of U.S. counterterrorism policy.

"The military is easily capable of adapting to change," the source told me, "but they don't like to stop anything they feel is making their lives easier or is to their benefit. And this certainly is, in their eyes, a very quick, clean way of doing things. It's a very slick, efficient way to conduct the war, without having to have the massive ground invasion mistakes of Iraq and Afghanistan. But at this point they have become so addicted to this machine, to this way of doing business, that it seems like it's going to become harder and harder to pull them away from it the longer they're allowed to continue operating in this way."

This book, most of which was originally published by *The Intercept* as "The Drone Papers,"[11] was produced by a team of reporters and researchers that spent months analyzing the documents and is intended to serve as a long-overdue public examination of the methods and outcomes of America's assassination program. That campaign, carried out by two presidents through four pres-

An MQ-1 Predator unmanned aircraft.

idential terms, has been conducted secretly. The public has a right to see these documents not only to engage in an informed debate about the future of U.S. wars, both overt and covert, but also to understand the circumstances under which the U.S. government arrogates to itself the right to sentence individuals to death without the established checks and balances of arrest, trial, and appeal.

Among the key revelations uncovered by *The Intercept* are the following.

HOW THE PRESIDENT AUTHORIZES TARGETS FOR ASSASSINATION.

It has been widely reported that President Obama directly approves high-value targets for inclusion on the kill list. The secret ISR study provides new insight into the kill chain, including a detailed chart stretching from electronic and human intelligence gathering all the way to the president's desk. In the same month the ISR study was circulated, May 2013, Obama signed the policy guidance on the use of force in counterterrorism operations overseas. A senior administration official, who declined to comment on the classified documents, admitted that "those guidelines remain in effect today."

As we detail in chapter 2, U.S. intelligence personnel collect information on potential targets drawn from government watchlists and the work of intelligence, military, and law enforcement agencies. At the time of the ISR study, when someone was destined for the kill list, intelligence analysts created a portrait of the suspect and the threat that person posed, pulling it together "in a condensed format known as a 'baseball card.'" That information was then bundled with operational information in a "target information folder" to be "staffed up to higher echelons" for action. On average, one slide indicates, it took fifty-eight days for the president to sign off on a target. At that point U.S. forces had sixty days to carry out the strike. The documents include two case studies that are partially based on information detailed on baseball cards.

The system for creating baseball cards and targeting packages, according to the source, depends largely on intelligence intercepts and a multilayered system of fallible, human interpretation. "It isn't a surefire method," he said. "You're relying on the fact that you do have all these very powerful machines, capable of collecting extraordinary amounts of data and information," which can lead personnel involved in targeted killings to believe they have "godlike powers."

ASSASSINATIONS DEPEND ON UNRELIABLE INTELLIGENCE AND DISRUPT INTELLIGENCE GATHERING.

In undeclared war zones the U.S. military has become overly reliant on signals intelligence, or SIGINT, to identify and ultimately hunt down and kill people. The documents acknowledge that using metadata from phones and computers, as well as communications intercepts, is an inferior method of finding and finishing targeted people. They describe SIGINT capabilities on these unconventional battlefields as "poor" and "limited." Yet such collection, much of it provided by foreign partners, accounted for more than half the intelligence used to track potential kills in Yemen and Somalia. The ISR study characterized these failings as a technical hindrance to efficient operations, omitting the fact that faulty intelligence has led to the killing of innocent people, including U.S. citizens, in drone strikes.[12]

The source underscored the unreliability of metadata, most often from phone and computer communications intercepts. These sources of information, identified by so-called selectors, such as a phone number or email address, are the primary tools used by the military to find, fix, and finish its targets. "It requires an enormous amount of faith in the technology that you're using," the source said. "There's countless instances where I've come across intelligence that was faulty." This, he said, is a primary factor in the killing of civilians. "It's stunning the number of instances when selectors are misattributed to certain people. And it isn't until several months or years later that you all of a sudden realize that the entire time you thought you were going after this really hot target, you wind up realizing it was his mother's phone the whole time."

The source described how members of the special operations community view the people being hunted by the United States for possible death by drone strike: "They have no rights. They have no dignity. They have no humanity to themselves. They're just a 'selector' to an analyst. You eventually get to a point in the target's life cycle that you are following them, you don't even refer to them by their actual name." This practice, he said, contributes to "dehumanizing the people before you've even encountered the moral question 'Is this a legitimate kill or not?'"

The authors of the ISR study admit that killing suspected terrorists, even if they are "legitimate" targets, further hampers intelligence gathering: "Kill operations significantly reduce the intelligence available." A chart shows that special operations actions in the Horn of Africa resulted in captures just 25 percent of the time, indicating a heavy tilt toward lethal strikes.

STRIKES OFTEN KILL MANY MORE THAN THE INTENDED TARGET.

The White House and Pentagon boast that the targeted killing program is precise and that numbers of civilian deaths are minimal. However, documents detailing a special operations campaign in northeastern Afghanistan, Operation Haymaker, show that between January 2012 and February 2013, U.S. special operations airstrikes killed more than two hundred people. Of those, only thirty-five were the intended targets. During one four-and-a-half-month period of the operation, according to the documents, nearly 90 percent of the people killed in airstrikes were not the intended targets. In Yemen and Somalia, where the United States has far more limited intelligence capabilities to confirm the people killed are the intended targets, the equivalent ratios may well be much worse.

"Anyone caught in the vicinity is guilty by association," the source said. "[When] a drone strike kills more than one person, there is no guarantee that those persons deserved their fate. . . . So it's a phenomenal gamble."

THE MILITARY LABELS UNKNOWN PEOPLE IT KILLS "ENEMIES KILLED IN ACTION."

The documents show that the military designated people it killed in targeted strikes as EKIA, "enemy killed in action," even if they were not the intended targets of the strike. Unless evidence posthumously emerged to prove the males killed were not terrorists or "unlawful enemy combatants," EKIA remained their designation, according to the source. That process, he said, "is insane. But we've made ourselves comfortable with that. The intelligence community, JSOC, the CIA, and everybody that helps support and prop up these programs, they're comfortable with that idea." The source described official U.S. government statements minimizing the number of civilian casualties inflicted by drone strikes as "exaggerating at best, if not outright lies."

THE NUMBER OF PEOPLE TARGETED FOR DRONE STRIKES AND OTHER FINISHING OPERATIONS.

According to one secret slide, as of June 2012, there were sixteen people in Yemen whom President Obama had authorized U.S. special operations forces to assassinate. In Somalia there were four. The statistics contained in the documents appear to refer only to targets approved under the 2001 Authorization for Use of Military Force, not CIA operations. In 2012 alone, according to data compiled by the Bureau of Investigative Journalism, there were more than two hundred people killed in operations in Yemen and between four and eight in Somalia.

HOW GEOGRAPHY SHAPES THE ASSASSINATION CAMPAIGN.

In Afghanistan and Iraq the pace of U.S. drone strikes was much quicker than in Yemen and Somalia. This appears due, in large part, to the fact that Afghanistan and Iraq were declared war zones, and in

Iraq the United States was able to launch attacks from bases closer to the targeted people. By contrast, in Somalia and Yemen, undeclared war zones where strikes were justified under tighter restrictions, U.S. attack planners described a serpentine bureaucracy for obtaining approval for assassination. The secret ISR study states that the number of high-value targeting operations in these countries was "significantly lower than previously seen in Iraq and Afghanistan" because of these "constraining factors."

Even after the president approved a target in Yemen or Somalia, the great distance between drone bases and targets created significant challenges for U.S. forces, a problem referred to in the documents as the "tyranny of distance." In Iraq more than 80 percent of "finishing operations" were conducted within 150 kilometers of an air base. In Yemen the average distance was about 450 kilometers, and in Somalia it was more than 1,000 kilometers. On average, one document states, it took the United States six years to develop a target in Somalia, but just 8.3 months to kill the target once the president had approved his addition to the kill list.

INCONSISTENCIES WITH WHITE HOUSE STATEMENTS ABOUT TARGETED KILLING.

The White House's publicly available policy standards state that lethal force will be launched only against targets who pose a "continuing, imminent threat to U.S. persons." In the documents, however, there is only one explicit mention of a specific criterion: that a person "presents a threat to U.S. interest or personnel." While such a criterion may make sense in the context of a declared war in which U.S. personnel are on the ground in large numbers, such as in Afghanistan, that standard is so vague as to be virtually meaningless in countries like Yemen and Somalia, where very few U.S. personnel operate.

While many of the documents provided to *The Intercept* contain explicit internal recommendations for improving unconventional U.S. warfare, the source said that what is implicit is even more significant: the mentality reflected in the documents on the assassination programs. "This process can work. We can work out the kinks. We can excuse the mistakes. And eventually we will get it down to the point where we don't have to continuously come back . . . and explain why a bunch of innocent people got killed."

The architects of what amounts to a global assassination campaign do not appear concerned with either its enduring impact or its moral implications. "All you have to do is take a look at the world and what it's become, and the ineptitude of our Congress, the power grab of the executive branch over the past decade," the source said. "It's never considered: Is what we're doing going to ensure the safety of our moral integrity? Of not just our moral integrity, but the lives and humanity of the people that are going to have to live with this the most?"

DECODING THE LANGUAGE OF COVERT WARFARE

JOSH BEGLEY

Woven throughout this book are sidebars defining terms used in the secret documents provided to *The Intercept*. Together they form a labyrinth with twelve entrances and no exit.

187←

BIRDS

→35

The first bomb that was dropped from an airplane exploded in an oasis outside Tripoli on November 1, 1911.*

While flying over Ain Zara, Libya, Lt. Giulio Gavotti leaned out of his airplane, which looked like a dragonfly, and dropped a Haasen hand grenade. It landed "in the camp of the enemy, with good results."

One hundred years later the bombing is done by pilotless planes. They are controlled remotely, often half a world away. We have come to call them "drones."

People on the inside call them "birds."

Operators can watch their targets for hours, often from air-conditioned rooms, until they receive the order to fire. When the time is right, a room full of people watch as the shot is taken.

This is where they sit.

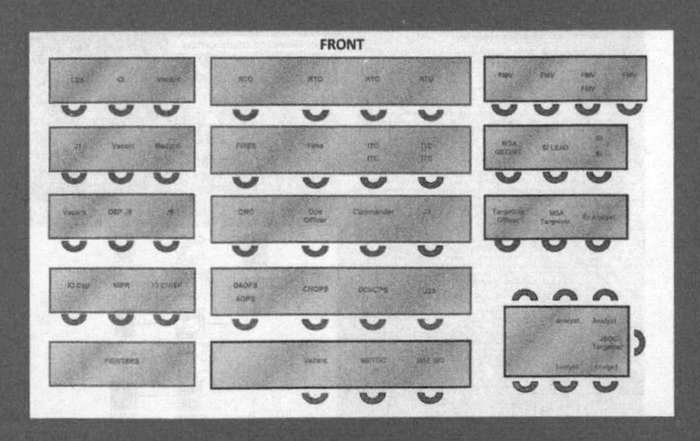

*With thanks to Sven Lindqvist's *A History of Bombing*, which served as a template for this narrative.

DEATH AND THE WATCHLIST

JEREMY SCAHILL
AND
RYAN DEVEREAUX

One
BARACK OBAMA
COMMANDER IN CHIEF

In 2013 the Obama administration quietly approved a substantial expansion of the terrorist watchlist system, authorizing a secret process that requires neither "concrete facts" nor "irrefutable evidence" to designate an American or foreigner as a terrorist. The "March 2013 Watchlisting Guidance," a 166-page document issued by the National Counterterrorism Center, spelled out the government's secret rules for putting individuals on its main terrorist database; the no-fly list, which prohibits air travel; and the selectee list, which triggers enhanced screening at airports and border crossings.[1]

Published by *The Intercept* in July 2014, the "Watchlisting Guidance" allows the government to designate individuals as representatives of terror organizations without any evidence they are actually connected to such organizations. It also gives a single White House official the unilateral authority to place entire "categories" of people the government is tracking onto the no-fly and selectee lists. The rules broaden the authority of government officials to "nominate" people to the watchlists based on what is vaguely described as "fragmentary information." They also allow for dead people to be watchlisted.

Over the years, the Bush and Obama administrations have fiercely resisted disclosing the criteria for placing names on the databases, though the guidelines are officially labeled as unclassified. In May

2014, Attorney General Eric Holder even invoked the state secrets privilege to prevent watchlisting guidelines from being disclosed in litigation launched by an American who was on the no-fly list. In an affidavit, Holder called them a "clear roadmap" to the government's terrorist-tracking apparatus, adding: "The Watchlisting Guidance, although unclassified, contains national security information that, if disclosed . . . could cause significant harm to national security."[2]

The guidelines were developed behind closed doors by representatives of the nation's intelligence, military, and law-enforcement establishment, including the Pentagon, CIA, National Security Agency (NSA), and FBI. Emblazoned with the crests of nineteen agencies, the document is the most complete and revealing look to date into the secret history of the government's terror list policies. It reveals a confounding and convoluted system filled with exceptions to its own rules, and it relies on the elastic concept of "reasonable suspicion" as a standard for determining whether someone is a possible threat. Because the government tracks "suspected terrorists" as well as "known terrorists," individuals can be watchlisted if they are suspected of being a suspected terrorist or if they are suspected of associating with people who are suspected of terrorism activity.

"Instead of a watchlist limited to actual, known terrorists, the government has built a vast system based on the unproven and flawed premise that it can predict if a person will commit a terrorist act in the future," said Hina Shamsi, the director of the National Security Project of the American Civil Liberties Union (ACLU). "On that dangerous theory, the government is secretly blacklisting people as suspected terrorists and giving them the impossible task of proving themselves innocent of a threat they haven't carried out." Shamsi, who reviewed the document, added, "These criteria should never have been kept secret."

The document's definition of "terrorist activity" includes actions that fall far short of bombing or hijacking. In addition to expected crimes, such as assassination and hostage-taking, the guidelines specify destruction of government property and damaging computers used by financial institutions as activities meriting placement on a list. They also define as terrorism any act that is "dangerous" to property and intended to influence government policy through intimidation.

This combination – a broad definition of what constitutes terrorism and a low threshold for designating someone a terrorist – opens the way to ensnaring innocent people in secret government dragnets. It can also be counterproductive. When resources are devoted to tracking people who are not genuine risks to national security, the actual threats get fewer resources – and might go unnoticed.

"If reasonable suspicion is the only standard you need to label somebody, then it's a slippery slope we're sliding down here, because then you can label anybody anything," said David Gomez, a former senior FBI special agent with experience running high-profile terrorism investigations. "Because you appear on a telephone list of somebody doesn't make you a terrorist. That's the kind of information that gets put in there."

The fallout is personal too. There are severe consequences for people unfairly labeled a terrorist by the U.S. government, which shares its watchlist data with local law enforcement, foreign governments, and "private entities." Once the U.S. government secretly labels you a terrorist or terrorist suspect, other institutions tend to treat you as one. It can become difficult to get a job or simply to stay out of jail. It can become burdensome – or impossible – to travel. And routine encounters with law enforcement can turn into ordeals.

In 2012 Tim Healy, the former director of the FBI's Terrorist Screening Center, described to CBS News how police officers use watchlists. "So if you are speeding, you get pulled over, they'll query that name," he said. "And if they are encountering a known or suspected terrorist, it will pop up and say call the Terrorist Screening Center. . . . So now the officer on the street knows he may be dealing with a known or suspected terrorist."[3] The problem is that the "known or suspected terrorist" might just be an ordinary citizen who should not be treated as a menace to public safety.

Until 2001 the government did not prioritize building a watchlist system; on 9/11 the government's list of people barred from flying included just sixteen names. By 2013, however, the no-fly list had swelled to tens of thousands of "known or suspected terrorists," or KSTs. The selectee list subjects people to extra scrutiny and questioning at airports and border crossings. The government has created several other databases, too. The largest is the Terrorist Identities Datamart Environment (TIDE), which gathers terrorism information from sensitive military and intelligence sources around the world. Because TIDE contains classified information that cannot be widely distributed, there is yet another list, the Terrorist Screening Database

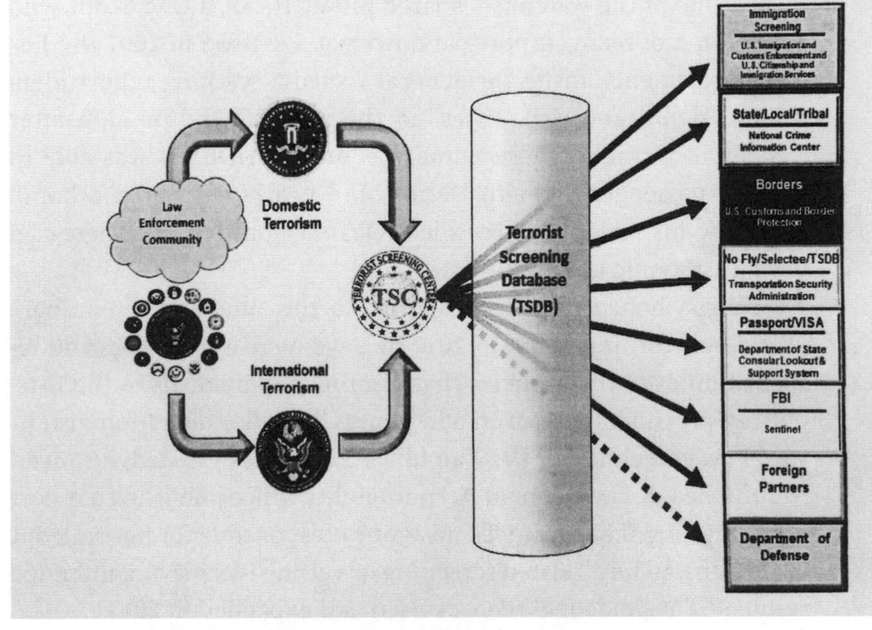

A chart from the "March 2013 Watchlisting Guidance."

(TSDB), which has been stripped of TIDE's classified data so that it can be shared. When government officials refer to "the watchlist," they are typically referring to the TSDB. (TIDE is the responsibility of the National Counterterrorism Center; the TSDB is managed by the Terrorist Screening Center at the FBI.)

In a statement a spokesman for the National Counterterrorism Center said that "the watchlisting system is an important part of our layered defense to protect the United States against future terrorist attacks" and that "watchlisting continues to mature to meet an evolving, diffuse threat." He added that U.S. citizens are afforded extra protections to guard against improper listing and that no one can be placed on a list solely for activities protected by the First Amendment. A representative of the Terrorist Screening Center did not respond to a request for comment.

The system has been criticized for years. In 2004 Senator Ted Kennedy complained that he was barred from boarding flights on five separate occasions because his name resembled the alias of a suspected terrorist.[4] Two years later CBS News obtained a copy of the no-fly list and reported that it included the names of Bolivian president Evo Morales and Nabih Berri, speaker of Lebanon's Parliament.[5] One of the watchlists snared Mikey Hicks, a Cub Scout who got his first of many airport pat-downs at age two.[6] In 2007 the Justice Department's inspector general issued a scathing report identifying "significant weaknesses" in the system.[7] And in 2009, after a Nigerian terrorist whose name was on the TIDE list was able to board a passenger flight to Detroit and nearly detonated a bomb sewn into his underwear, President Obama admitted that there had been a "systemic failure."[8]

Obama hoped that his response to the "underwear bomber" would be a turning point. In 2010 he gave increased powers and responsibilities to the agencies that nominate individuals to the lists, putting pressure on them to add names. His administration also issued new guidelines for the watchlists. Problems persisted, however. In 2012 the U.S. Government Accountability Office published a report clearly stating that there was no agency responsible for figuring out "whether watchlist-related screening or vetting is achieving intended results."[9] The guidelines were revised and expanded in 2013.

The five chapters and eleven appendixes of the "Watchlisting Guidance" are filled with acronyms, legal citations, and numbered paragraphs; it reads like an arcane textbook with a vocabulary all its own. Different types of data on suspected terrorists are referred to as "derogatory information," "substantive derogatory information," "extreme derogatory information," and "particularized derogatory information." The names of suspected terrorists are passed along a bureaucratic ecosystem of "originators," "nominators," "aggregators," "screeners," and "encountering agencies." And "upgrade," usually a happy word for travelers, is repurposed to mean that an individual has been placed on a more restrictive list.

The heart of the document revolves around the rules for placing individuals on a watchlist. "All executive departments and agencies" are responsible for collecting and sharing information on terrorist suspects with the National Counterterrorism Center. It sets a low standard – "reasonable suspicion" – for placing names on the watchlists and offers a multitude of vague, confusing, or contradictory instructions for gauging what is "reasonable." In the chapter titled "Minimum Substantive Derogatory Criteria" – even the title is hard to digest – the key sentence on reasonable suspicion offers little clarity: "To meet the REASONABLE SUSPICION standard, the NOMINATOR, based on the totality of the circumstances, must rely upon articulable intelligence or information which, taken together with rational inferences from those facts, reasonably warrants a determination that an individual is known or suspected to be or has been knowingly engaged in conduct constituting, in preparation for, in aid of, or related to TERRORISM and/or TERRORIST ACTIVITIES."

The "Watchlisting Guidance" makes no effort to define "articulable intelligence or information," an essential phrase in the passage. After stressing that hunches do not qualify as reasonable suspicion and that "there must be an objective factual basis" for labeling someone a terrorist, it goes on to state that no actual facts are required: "In determining whether a REASONABLE SUSPICION exists, due weight should be given to the specific reasonable inferences that a NOMINATOR is entitled to draw from the facts in light of his/her experience and not on unfounded suspicions or hunches. Although irrefutable evidence or concrete facts are not necessary, to be reasonable, suspicion should be as clear and as fully developed as circumstances permit."

While the guidelines nominally prohibit nominations based on unreliable information, they explicitly regard "uncorroborated" Facebook or Twitter posts as sufficient grounds for putting an individual on one of the watchlists: "Single source information, including but not limited to 'walk-in,' 'write-in,' or postings on social media sites, however, should not automatically be discounted. . . . The NOMINATING AGENCY should evaluate the credibility of the source, as well as the nature and specificity of the information, and nominate even if that source is uncorroborated."

There are a number of loopholes for putting people onto the watchlists even if the criterion of reasonable suspicion cannot be met. One is clearly defined: the immediate family of suspected terrorists – spouse, children, parents, and siblings – may be watchlisted without any suspicion that they themselves are engaged in terrorist activity. But another loophole is quite broad: "associates" who have a defined relationship with a suspected terrorist but whose involvement in terrorist activity is not known. A third loophole is broader still: individuals with "a possible nexus" to terrorism but for whom there is not enough "derogatory information" to meet the reasonable suspicion standard.

Americans and foreigners can be nominated for the watchlists if they are associated with a terrorist group, even if that group has not been designated a terrorist organization by the U.S. government. They can also be treated as "representatives" of a terrorist group even if they have "neither membership in nor association with the organization." The guidelines do helpfully note that certain associations, such as providing janitorial services or delivering packages, are not grounds for being watchlisted.

The nomination system appears to lack meaningful checks and balances. Although government officials have repeatedly said there is a rigorous process for making sure no one is unfairly placed in the databases, the guidelines acknowledge that all nominations of "known terrorists" are considered justified unless the National Counterterrorism Center has evidence to the contrary. In an April 2014 court filing, the government disclosed that there were 468,749 KST nominations in 2013, of which only 4,915 were rejected – a rate of about 1 percent.[10] The rules appear to invert the legal principle of due process, defining nominations as "presumptively valid."

While on paper the nomination process appears to be methodical, in practice there is a shortcut around the entire system. Known as a "threat-based expedited upgrade," it gives a single White House official the unilateral authority to elevate entire "categories of people" whose names appear in the larger databases onto the no-fly or selectee lists. This can occur, the "Watchlisting Guidance" states, when there is a "particular threat stream" indicating that a certain type of individual may commit a terrorist act. This extraordinary power for "categorical watchlisting" – otherwise known as profiling – is vested in the assistant to the president for homeland security and counterterrorism, a position formerly held by CIA director John Brennan, that does not require Senate confirmation.

The "Watchlisting Guidance" does not indicate which "categories of people" have been subjected to threat-based upgrades. It is not clear, for example, whether a category might be as broad as military-age males from Yemen. The guidelines do make clear that American citizens and green-card holders are subject to such upgrades, though government officials are required to review their status in an "expedited" procedure. Upgrades can remain in effect for seventy-two hours before being reviewed by a small committee of senior officials. If approved, they can remain in place for thirty days before a renewal is required and can continue "until the threat no longer exists."

"In a set of watchlisting criteria riddled with exceptions that swallow rules, this exception is perhaps the most expansive and certainly one of the most troubling," said Shamsi, the ACLU attorney. "It's reminiscent of the Bush administration's heavily criticized color-coded threat alerts, except that here, bureaucrats can exercise virtually standard-less authority in secret with specific negative consequences for entire categories of people." The National Counterterrorism Center declined to provide any details on the upgrade authority, including how often it has been exercised and for what categories of people.

The "Watchlisting Guidance" provides the clearest explanation yet of what is happening when Americans and foreigners are pulled aside at airports and border crossings by government agents. The fifth chapter, titled "Encounter Management and Analysis," details the type of information that is targeted for collection during "encounters" with people on the watchlists, as well as the different organizations that should collect the data. The Department of Homeland Security is described as having the largest number of encounters, but other authorities, ranging from the State Department and Coast Guard to foreign governments and "certain private entities," are also involved in assembling "encounter packages" when watchlisted individuals cross their paths. The encounters can be face-to-face meetings or electronic interactions, for instance, when a watchlisted individual applies for a visa.

In addition to data like fingerprints, travel itineraries, identification documents, and gun licenses, the rules encourage screeners to acquire health insurance information, drug prescriptions, "any cards with an electronic strip on it (hotel cards, grocery cards, gift cards, frequent flyer cards)," cell phones, email addresses, binoculars, peroxide, bank account numbers, pay stubs, academic transcripts, parking and speeding tickets, and want ads. The digital information singled out for collection includes social media accounts, cell phone lists, speed-dial numbers, laptop images, thumb drives, iPods, Kindles, and cameras. All of the information is then uploaded to the TIDE database.

Screeners are also instructed to collect data on any "pocket litter," scuba gear, EZ Passes, library cards, and book titles, along with information about their condition, "e.g., new, dog-eared, annotated, unopened." Business cards and conference materials are also targeted, as well as "anything with an account number" and information about any gold or jewelry worn by the watchlisted individual. Even "animal information" – details about pets from veterinarians or tracking chips – is requested. The guidelines also encourage the collection of biometric and biographical data about the travel partners of watchlisted individuals.

The list of government entities that collect this data includes the U.S. Agency for International Development, which is neither an intelligence nor a law-enforcement agency but, as the guidelines note, funds foreign aid programs that promote environmentalism, health care, and education. USAID, which presents itself as committed to fighting global poverty, nonetheless appears to serve as a conduit for sensitive intelligence about foreigners. According to the "Watchlisting Guidance," "When USAID receives an application seeking financial assistance, prior to granting, these applications are subject to vetting by USAID intelligence analysts at the [Terrorist Screening Center]." The guidelines do not disclose the volume of names provided by USAID, the type of information it provides, or the number and duties of the "USAID intelligence analysts."

A USAID spokesman said that "in certain high-risk countries, such as Afghanistan, USAID has determined that vetting potential partner organizations with the terrorist watchlist is warranted to protect U.S. taxpayer dollars and to minimize the risk of inadvertent funding of terrorism." He stated that since 2007, the agency has checked "the names and other personal identifying information of key individuals of contractors and grantees, and sub-recipients."

The government has been widely criticized for making it impossible for people to know why they have been placed on a watchlist and for making it nearly impossible to get off. The guidelines state that "the general policy of the U.S. Government is to neither confirm nor deny an individual's watchlist status." But the courts have taken exception to the official silence and foot-dragging: in June 2014 a federal judge described the government's secretive removal process as unconstitutional and "wholly ineffective."[11]

The difficulty of getting off the list is highlighted by a passage in the "Watchlisting Guidance" stating that an individual can be kept on the watchlist, or even placed onto the watchlist, despite being acquitted of a terrorism-related crime. The guidelines justify this by noting that conviction in U.S. courts requires evidence beyond a reasonable doubt, whereas watchlisting requires only a reasonable suspicion. Once suspicion is raised, even a jury's verdict cannot erase it.

Not even death provides a guarantee of getting off the list. The guidelines state that the names of dead people will stay on the list if there is reason to believe the deceased's identity may be used by a suspected terrorist, which the National Counterterrorism Center calls a "demonstrated terrorist tactic." For the same reason, the rules permit the names of deceased spouses of living suspected terrorists to be placed on the list.

For the living, the process of getting off the watchlist is simple yet opaque. A complaint can be filed with the Department of Homeland Security's Traveler Redress Inquiry Program, which will launch an internal review that is not subject to oversight by any court or entity outside the counterterrorism community. The review may result in removal from a watchlist or an adjustment of watchlist status, but the individual will not be told if he or she prevails. The guidelines highlight one of the reasons it has been difficult to get off the list: when multiple agencies have contributed information on a watchlisted individual, all of them must agree to removing that person.

If U.S. citizens are placed on the no-fly list while abroad and are turned away from a flight bound for the United States, the guidelines explain that they should be referred to the nearest U.S. embassy or consulate, which is prohibited from informing them why they were blocked from flying. These individuals can be granted a "one-time waiver" to fly, though they will not be told that they are traveling on a waiver. Back in the United States they will be prevented from boarding another flight. Nominating agencies are "under a continuing obligation" to provide exculpatory information when it emerges and are expected to conduct annual reviews of watchlisted American citizens and green card holders. It is unclear whether foreigners – or the dead – are reviewed at the same pace. As the guidelines note, "watchlisting is not an exact science."

In fact, according to another set of classified government documents, published in August 2014 by *The Intercept*, nearly half of the people on the U.S. government's widely shared database of terrorist suspects were not connected to any known terrorist group.[12] Of the 680,000 people caught up in the government's Terrorist Screening Database, a watchlist of "known or suspected terrorists" that is shared with local law enforcement agencies, private contractors, and foreign governments, more than 40 percent were described by the government as having "no recognized terrorist group affiliation." That group – 280,000 people – dwarfed the number of watchlisted people suspected of ties to al Qaeda, Hamas, and Hezbollah combined.

The documents, obtained from a source in the intelligence community, reveal that the Obama administration has presided over an unprecedented expansion of the terrorist screening system. Since taking office Obama has boosted the number of people on the no-fly list more than tenfold, to an all-time high of 47,000, surpassing the number of people barred from flying under George W. Bush.

"If everything is terrorism, then nothing is terrorism," said David Gomez, a former senior FBI special agent. The watchlisting system, he added, is "revving out of control."

The classified documents were prepared by the National Counterterrorism Center, the lead agency for tracking individuals with suspected links to international terrorism. Stamped "SECRET" and "NOFORN" (indicating they are not to be shared with foreign governments), they offer the most complete numerical picture of the watchlisting system to date. The documents reveal that the government adds names to its databases, or adds information on existing subjects, at a rate of nine hundred records each day. They also provide evidence of blanket ethnic profiling: the second-highest concentration of people designated as "known or suspected terrorists" by the government was in Dearborn, Michigan, a city of 96,000 that has the largest percentage of Arab American residents in the country. The documents also reveal that the CIA uses a previously unknown program, code-named Hydra, to secretly access databases maintained by foreign countries and extract data to add to the watchlists.

A U.S. counterterrorism official familiar with watchlisting data said that as of November 2013, there were approximately 700,000 people in the Terrorist Screening Database, or TSDB, but he declined

WHO'S ON THE WATCHLIST

NO RECOGNIZED TERRORIST GROUP AFFILIATION (280,000)

"OTHER" RECOGNIZED TERRORIST GROUPS (92,765)

AL QAEDA IN IRAQ (73,189)

TALIBAN (62,794)

AL QAEDA (50,446)

HAMAS (21,913)

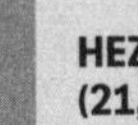

HEZBOLLAH (21,199)

HAQQANI NETWORK (12,491)

AL SHABAB (11,547)

FARC (11,275)

ASA'IB AHL AL-HAQ (8,385)

AL QAEDA IN THE ARABIAN PENINSULA (8,211)

All figures taken from an August 2013 slide produced by the National Counterterrorism Center.

to provide current numbers. In July 2014 the Associated Press, citing federal court filings by government lawyers, reported that 1.5 million names had been added to the watchlist over the previous five years.[13] The government official called that a misinterpretation of the data: "The list has grown somewhat since that time but is nowhere near the 1.5 million figure cited in recent news reports." He added that the statistics cited by the Associated Press include not just nominations of individuals but also bits of intelligence and biographical information obtained on watchlisted persons.

When U.S. officials refer to "the watchlist," they typically mean the TSDB, an unclassified pool of information shared across the intelligence community and the military, as well as local law enforcement, foreign governments, and private contractors. According to the government's March 2013 watchlisting guidelines, officials don't need "concrete facts" or "irrefutable evidence" to secretly place someone on the list – only a vague and elastic standard of "reasonable suspicion."

"You need some fact-basis to say a guy is a terrorist, that you know to a probable-cause standard that he is a terrorist," said Gomez, the former FBI agent. "Then I say, 'Build as big a file as you can on him.' But if you just suspect that somebody is a terrorist? Not so much."

The National Counterterrorism Center did not respond to questions about its terrorist screening system. Instead, in a statement, it praised the watchlisting system as a "critical layer in our counterterrorism defenses" and described it as superior to the pre-9/11 process for tracking threats, which relied on lists that were "typed or hand-written in card catalogues and ledgers." The White House declined to comment.

Most people placed on the government's watchlist begin in the larger – and more invasive – classified Terrorist Identities Datamart Environment. The TIDE database actually allows for targeting people based on far less evidence than the already lax standards used for placing people on the watchlist. TIDE's information is shared across the U.S. intelligence community, as well as with commando units from the Special Operations Command and with domestic agencies such as the New York City Police Department. In the summer of 2013 officials celebrated what one classified document prepared by the National Counterterrorism Center refers to as "a milestone": boosting the total number of people in the TIDE database to one million, up from half a million four years earlier.

The document credits that historic achievement to the Directorate of Terrorist Identities (DTI), a secretive and virtually unknown U.S. counterterrorism unit responsible for maintaining TIDE: "This number is a testament to DTI's hard work and dedication over the past 2.5 years." The number is also a testament to the Obama administration's intensified collection of personal information on individuals with suspected links to terrorism. In 2006, when CBS News obtained a copy of the no-fly list, it included 44,000 names. Faced with widespread public backlash, the government cut the list down to just 4,000 names by late 2009.

The next year, after the so-called underwear bomber tried to bring down a commercial airliner bound for Detroit, Obama loosened the criteria for adding people to the no-fly list. The impact was immediate. According to one document, since 2010, when Obama loosened the criteria for adding people to the no-fly list, the National Counterterrorism Center has "created more than 430,000 terrorism-related person records" while deleting the records of only 50,000 people "whose nexus to terrorism was refuted or did not meet current watchlisting criteria." The documents reveal that at that time more than 240 TIDE "nominations" were being processed each day.

"You might as well have a blue wand and just pretend there's magic in it, because that's what we're doing with this – pretending that it works," said former FBI agent Michael German, now a fellow at New York University's Brennan Center for Justice. "These agencies see terrorism as a winning card for them. They get more resources. They know that they can wave that card around and the American public will be very afraid and Congress and the courts will allow them to get away with whatever they're doing under the national security umbrella."

The documents emphasize that the government seeks to add only as many people to the TIDE list "as are necessary for our nation's counterterrorism mission." With hundreds of new nominations coming in every day, the numbers provide only a momentary snapshot of a watchlist system that is in constant motion.

An August 2013 slide from the National Counterterrorism Center titled "TIDE by the Numbers" lays out the scope of the Obama administration's watchlisting system and those it is targeting. According to the slide, which notes that the numbers are "approximate," worldwide 680,000 people have been watchlisted, with another 320,000 monitored in the larger TIDE database. As of August 2013, 5,000 Americans were on the watchlist while another 15,800 were targeted in TIDE. Another 16,000 people, including 1,200 Americans, had been classified as "selectees" targeted for enhanced screenings at airports and border crossings; 611,000 men and 39,000 women were on the main terrorist watchlist.

The top "nominating agencies" responsible for placing people on the government's watchlists were the CIA, the Defense Intelligence Agency, the National Security Agency, and the FBI. The top five U.S. cities represented on the main watchlist for "known or suspected terrorists" were New York; Dearborn, Michigan; Houston; San Diego; and Chicago. The inclusion of Dearborn confirms what residents and civil liberties advocates have frequently argued: that the Muslim, Arab, and Sikh communities in and around Dearborn are unfairly targeted by invasive law enforcement probes, unlawful profiling, and racism. "To my knowledge, there have been no Muslims in Dearborn who have committed acts of terrorism against our country," said Dawud Walid, executive director of the Michigan chapter of the Council on American-Islamic Relations. Walid added that the high concentration of Dearborn residents in the watchlisting system "just confirms the type of engagement the government has with our community—as seeing us as perpetual suspects."

The documents also list groups the government is targeting in its counterterrorism mission. The groups with the largest number of targeted people on the main terrorism watchlist—aside from "no recognized terrorist group affiliation"—were al Qaeda in Iraq (73,189), the Taliban (62,794), and al Qaeda (50,446), followed by Hamas (21,913) and Hezbollah (21,199). Although the Obama administration had repeatedly asserted that al Qaeda in the Arabian Peninsula posed the most significant external terrorist threat to the United States, the 8,211 people identified as being tied to the group actually represented the smallest category on the list of the top ten recognized terrorist organizations. AQAP was outnumbered by people suspected of ties to the Pakistan-based Haqqani Network (12,491), the Colombia-based FARC (11,275), and the Somalia-based al Shabaab (11,547).

The documents also reveal that as of 2013, the United States had designated 3,200 people as "known or suspected terrorists" associated

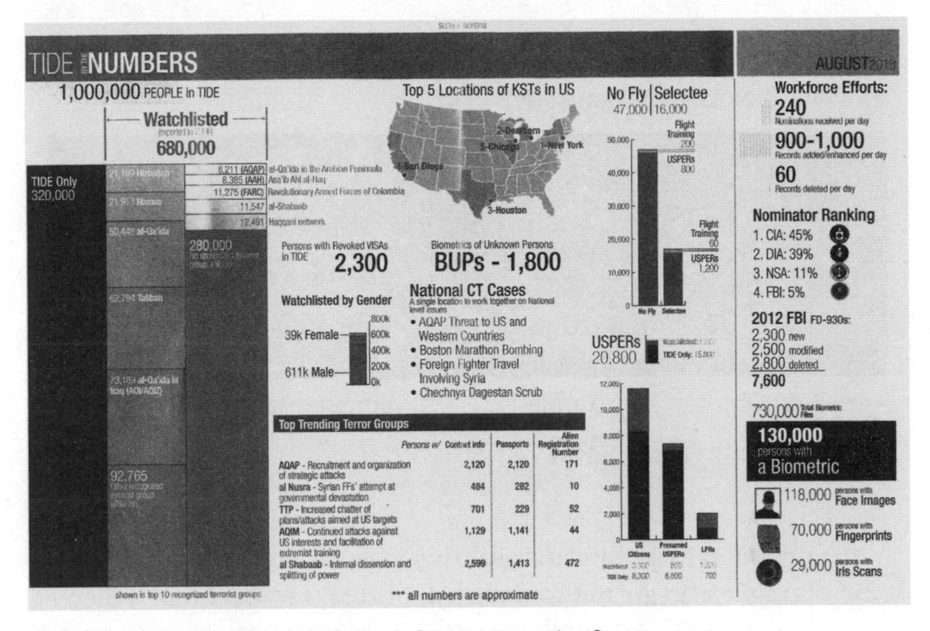

An August 2013 slide from the National Counterterrorism Center.

with the war in Syria. Among them were 715 Europeans and Canadians, as well as 41 Americans. In March 2014 Matt Olsen, the director of the National Counterterrorism Center, claimed that there were more than 12,000 foreign fighters in Syria, including more than 1,000 Westerners and roughly 100 Americans.

Clearly the government does much more than simply stop watchlisted people at airports. It also covertly collects and analyzes a wide range of personal information about those individuals, including facial images, fingerprints, and iris scans. In the aftermath of the Boston Marathon bombing, for instance, the Directorate of Terrorist Identities began an aggressive program to collect biometric data and other information on all Americans on the TIDE list: "This project includes record by record research of each person in relevant Department of State and [intelligence community] databases, as well as bulk data requests for information." The DTI also worked on the subsequent Chicago Marathon, performing "deep dives" for biometric and other data on people in the Midwest whose names were on the TIDE list. In the process the directorate pulled the TIDE records of every person with an Illinois, Indiana, or Wisconsin driver's license.

DTI's efforts in Boston and Chicago are part of a broader push to obtain biometric information on the more than one million people

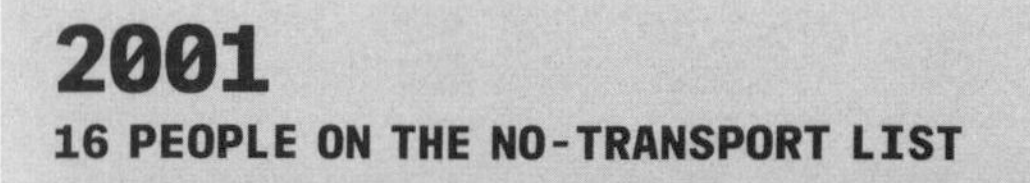

targeted in its secret database. This includes hundreds of thousands of people who are not watchlisted. In 2013 the directorate's Biometric Analysis Branch (BAB) launched an initiative to obtain biometric data from driver's license records across the country. At least fifteen states and the District of Columbia are working with the directorate to facilitate access to facial images from driver's licenses. In fiscal year 2013, 2,400 such images were provided for inclusion in the secret TIDE database. The Biometric Analysis Branch also offers its "unique skill of facial identification support" to a "broad customer base." In 2013 its analysts produced more than 290 reports for other government entities, including the CIA, the New York City Police Department, and the military's elite Special Operations Command.

All told, the classified documents show that the government compiles strikingly detailed dossiers of data on individuals who have been swept up in its databases. Though some of the documents offer conflicting information on how much biometric data the government collects, the most detailed report shows that the numbers are very significant. In 2013, for instance, the main terrorism database included more than 860,000 biometric files on 144,000 people. The database also contained more than half a million facial images, nearly a quarter of a million fingerprints, and 70,000 iris scans. The government also maintains biometric data on people that it hasn't identified; TIDE contained 1,800 "BUPs," or "biometrics of unknown persons." In a single year, moreover, the government expanded its

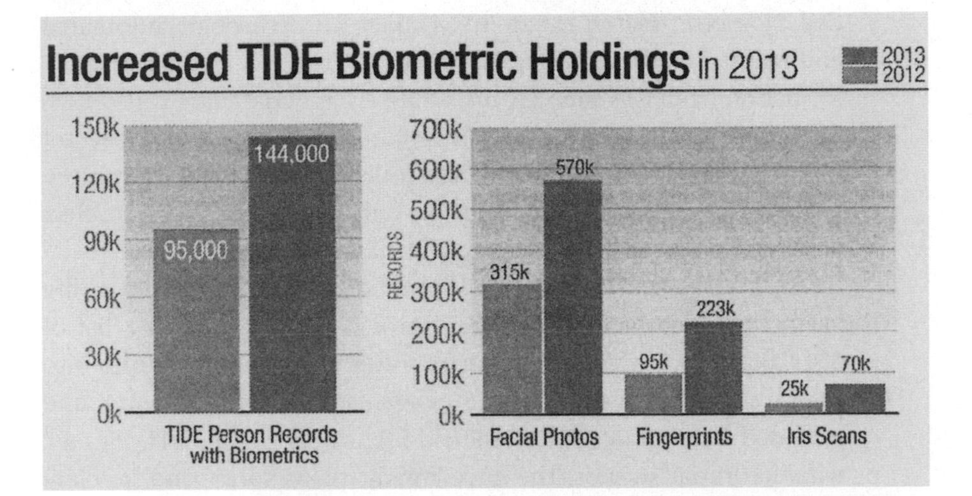

Two charts from "Directorate of Terrorist Identities (DTI): Strategic Accomplishments 2013."

collection of "non-traditional" biometric data, including dramatic increases in handwriting samples (32 percent), signatures (52 percent), scars, marks, and tattoos (70 percent), and DNA strands (90 percent).

"We're getting into Minority Report territory, when being friends with the wrong person can mean the government puts you in a database and adds DMV photos, iris scans, and face recognition technology to track you secretly and without your knowledge," said Hina Shamsi, director of the American Civil Liberties Union's National Security Project. The fact that this information can be shared with agencies from the CIA to the NYPD, which are not known for protecting civil liberties, brings us closer to an invasive and rights-violating government surveillance society at home and abroad."

The DTI also goes far beyond accessing information from state driver's licenses. In managing the main terrorism database, the directorate coordinates with the CIA and the National Media Exploitation Center, a Pentagon wing responsible for analyzing and disseminating "paper documents, electronic media, videotapes, audiotapes, and electronic equipment" seized abroad in military or intelligence operations. By sharing information with the military, the National Counterterrorism Center asserts, the DTI is able to "obtain additional data fusion points by accessing and exploiting NMEC data holdings." In return the directorate "provides NMEC with a classified biometric search capability against TIDE through automated and manual facial identification support." The DTI also harvests information from CIA sources, including a secret database called CINEMA, short for CIA Information Needs Management, and a secret CIA program called "Hydra," which utilizes "clandestinely acquired foreign government information" to enhance the quality of "select populations" in TIDE.

In 2013 DTI and the CIA ran a "proof of concept" for Hydra, using Pakistan as a guinea pig. The DTI provided the CIA with a list of 555 Pakistanis in the TIDE database. After inputting the names into Hydra, the CIA "vetted these names against Pakistani passports" and provided biographic and biometric identifiers to the DTI. Pleased with its initial success, the government plans to expand its clandestine data-mining operation. "Future initiatives," the documents note, "will include additional targeted countries."

13←

OBJECTIVES

→39

Most of the time drone operators are trying to kill someone specific. They call these people, the people being hunted, "objectives."

What does an objective look like? Here's an example.

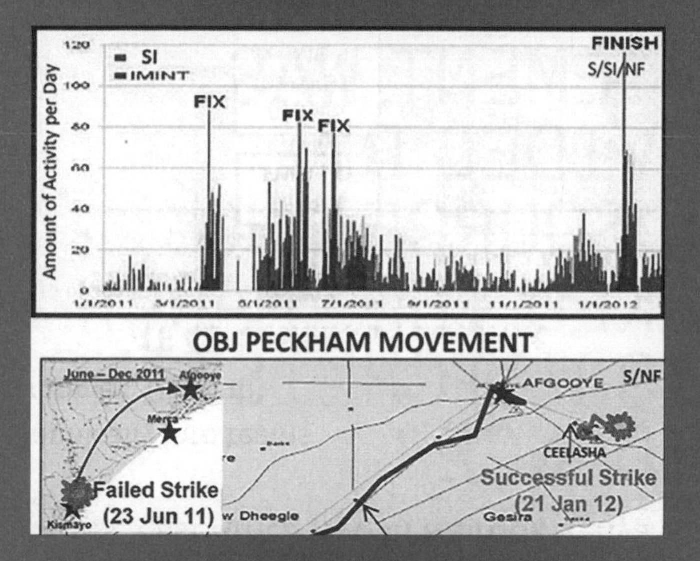

This timeline was for a man named Bilal el-Berjawi. Intelligence agencies watched him for years, then the British government stripped him of his citizenship.

After calling his wife, who had just given birth in a London hospital, Berjawi was killed by an American drone strike. Some people thought the call might have given away his location, but the drones already knew where he was.

This was his car.

WHY I LEAKED THE WATCHLIST DOCUMENTS

THE FOLLOWING STATEMENT WAS PROVIDED BY THE SOURCE WHO LEAKED THE "MARCH 2013 WATCHLISTING GUIDANCE"

Over the past few years we've heard a lot about President Obama's "secret kill list," yet still we know virtually nothing about its implementation. Despite mild congressional scrutiny and ACLU lawsuits directed at the shroud of secrecy, some basic questions remain unanswered:

How do you get on the list?
Am I on the list?
Who put me on the list?
How do you get off the list?
Can you get off the list?

The truth is, there are several such lists, used to target individuals for different reasons. Some lists are closely kept; others span multiple intelligence and local law enforcement agencies. There are lists used to kill or capture supposed "high-value targets," and others are intended to threaten, coerce, or simply monitor a person's activity. However, all the lists, whether to kill or to silence, originate from the Terrorist Identities Datamart Environment, and they are maintained by the Terrorist Screening Center at the National Counterterrorism Center. The existence of TIDE is unclassified, yet details about how it functions in our government are completely unknown to the public. In August 2013 the database reached a milestone of one million entries. Today it is thousands of entries larger and is growing faster than it has since its inception in 2003.

The "March 2013 Watchlisting Guidance" lays out the broad criteria for nominating someone to the database. Not only does the Terrorist Screening Center reserve the right to store your name, date of birth, and other basic identifying information, but it also stores your medical records, transcripts, and passport data; your license plate numbers, email, and cell-phone number (along with the phone's International Mobile Subscriber Identity and International Mobile Station Equipment Identity numbers); your bank account numbers and purchases; and other sensitive information, including DNA and photographs capable of identifying you using facial recognition software. The National Counterterrorism Center collaborates annually with agencies from the international alliance known as Five Eyes to supplement any information missing from entries already in its database or to add more entries. Individual entries in the database are assigned a TIDE personnel number, or TPN. From Osama bin Laden (TPN 1063599) to Abdulrahman Awlaki (TPN 26350617), the American son of Anwar al Awlaki, anyone who has ever been the target of a covert operation was first assigned a TPN and closely monitored by all agencies who follow that TPN long before they were eventually put on a separate list and extrajudicially sentenced to death.

When governments begin to tally enormous enemies lists, they run roughshod over our essential checks on power, especially when they consider their own citizens to be a threat. Of the more than one million entries in the TIDE database, approximately 21,000 are those of American citizens. By leaking this information (which is unclassified but, by reverse logic, considered too sensitive to be released), I hope to give the public an opportunity to know what kind of activity might lead to their being placed on a list used to monitor their everyday activity. For the first time the public has an opportunity to gain insight into the criteria that could potentially lead to their own trial by drone strike.

In 2008 I shook hands with Senator Obama when he came

through my town on his way to the White House. After his inauguration he said, "Transparency and the rule of law will be the touchstones of this presidency." I firmly believe those principles are crucial to an open society, which is why I was compelled to reveal this information. If this administration lacks the courage to uphold its promises to the people, then I and others like me will do so for them.

35←

JACKPOT

→53

When drone operators hit their target, killing the person they intend to kill, that person is called a "jackpot."

When operators miss their target and end up killing someone else, they label that person EKIA, or "enemy killed in action."

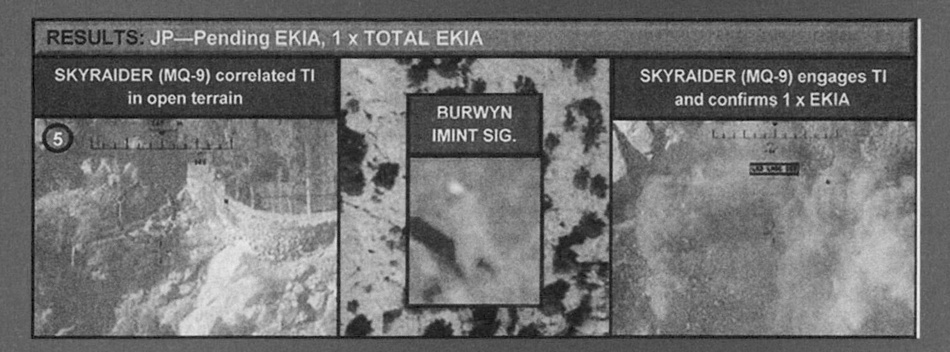

FIND, FIX,
FINISH
JEREMY SCAHILL

Soon after he was elected president, Barack Obama was strongly urged by Michael Hayden, the outgoing CIA director, and his new top counterterrorism adviser, John Brennan, to adopt the way of the scalpel: small-footprint counterterrorism operations and drone strikes. In one briefing Hayden told Obama that covert action was the only way to confront al Qaeda and other terrorist groups plotting attacks against the United States.

The view among Obama's inner circle was that Iraq and Afghanistan had served as useful laboratories for such tactics, but deploying them outside conventional war zones meant that different legal and diplomatic considerations would apply. An all-star team of special operations commanders, war planners, and Pentagon officials pressed the new president to dramatically ramp up the shadow wars in Yemen and Somalia to fight the emerging threats in those countries. They called for sweeping away bureaucratic obstacles and streamlining lethal operations. In short, a new global architecture of assassination was called for, and that meant navigating an increasingly tense turf war between the CIA and the Pentagon over these activities.

The CIA had long dominated the covert war in Pakistan, and in 2009 Obama expanded the agency's drone resources there and in Afghanistan to regularly pound al Qaeda, the Pakistani Taliban, and other targets.[1] The military, tasked with prosecuting the broader war in Afghanistan, was largely sidelined in the Pakistan theater, save for the occasional cross-border raid and the U.S. Air Force personnel who operated the CIA's drones.[2] But the Pentagon was not content to play a peripheral role in the global drone war and aggressively positioned itself to lead the developing drone campaigns in Yemen and Somalia.

In September 2009 Gen. David Petraeus, commander of U.S. Central Command (Centcom), issued a Joint Unconventional Warfare Task Force Execute Order that would lay the groundwork for military forces to conduct expanded clandestine actions in Yemen and other countries.[3] It allowed for U.S. special operations forces to enter friendly and unfriendly countries "to build networks that could 'penetrate, disrupt, defeat or destroy' al Qaeda and other militant groups, as well as to 'prepare the environment' for future attacks by American or local military forces." At the same time, both al Qaeda in the Arabian Peninsula and al Shabaab began to escalate their rhetoric and, in the case of al Qaeda in the Arabian Peninsula (AQAP), to plot terror attacks on U.S. soil. After the failed Christmas Day plot by the "underwear bomber," the Obama administration responded by green-lighting special operations commanders' plans for direct action.[4]

The insignia of the Joint Special Operations Command.

In December 2009 the Obama administration signed off on its first covert airstrike in Yemen, a cruise missile attack that killed more than forty people, most of them women and children.[5] After that strike, as with the CIA's program in Pakistan, drones would fuel the Joint Special Operations Command's high-value targeting campaign in the region. When Obama took office, there had been only one U.S. drone strike in Yemen, in November 2002.[6] By 2012 a drone strike was reported in Yemen every six days. As of August 2015, more than 490 people had been killed in drone strikes in Yemen alone.

"The drone campaign right now really is only about killing. When you hear the phrase 'capture/kill,' 'capture' is actually a misnomer. In the drone strategy that we have, 'capture' is a lowercase c. We don't capture people anymore," Lt. Gen. Michael Flynn, former head of the Defense Intelligence Agency, told me. "Our entire Middle East policy seems to be based on firing drones. That's what this administration decided to do in its counterterrorism campaign. They're enamored by the ability of special operations and the CIA to find a guy in the middle of the desert in some shitty little village and drop a bomb on his head and kill him."

The tip of the spear in the Obama administration's escalated wars in East Africa and the Arabian Peninsula was the special operations task force known as TF 48-4. Its primary command center was at the former French Army outpost at Camp Lemonnier in Djibouti, a small African nation nestled among Ethiopia, Eritrea, Somalia, and the Gulf of Aden. With its strategic location Lemonnier served as the hub for launching actions from military facilities scattered across the region. The task force also utilized a maritime drone platform and a surveillance apparatus positioned in the Arabian Sea, used for intercepting data. TF 48-4 had sites in Nairobi and Sanaa and a drone base in Arba Minch, Ethiopia. A small base in Manda Bay, Kenya – a stone's throw from Somalia – housed special operations commandos and manned aircraft.

The task force's operations, aimed at hunting down and killing or capturing members of al Qaeda in the Arabian Peninsula and al Shabaab, were largely conducted with drones and fixed-wing aircraft. On occasion small teams of special operators mounted ground operations inside Somalia and Yemen or interdicted ships, snatching suspected terrorists.[7] But drones were the administration's preferred weapon. "It is the politically advantageous thing to do – low cost, no U.S. casualties, gives the appearance of toughness," said Adm. Dennis Blair, Obama's former director of national intelligence, explaining how the administration viewed its policy at the time. "It plays well domestically, and it is unpopular only in other countries. Any damage it does to the national interest only shows up over the long term."[8]

As Yemen's status began to rise to the top of U.S. counterterrorism priorities, the long-simmering turf war between the Pentagon and the CIA flared up. In 2011 the CIA began using a newly constructed drone base in Saudi Arabia, giving it easier access to targets in Yemen than had the military's bases in East Africa. There were parallel, and competing, target lists and infighting over who should

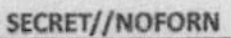

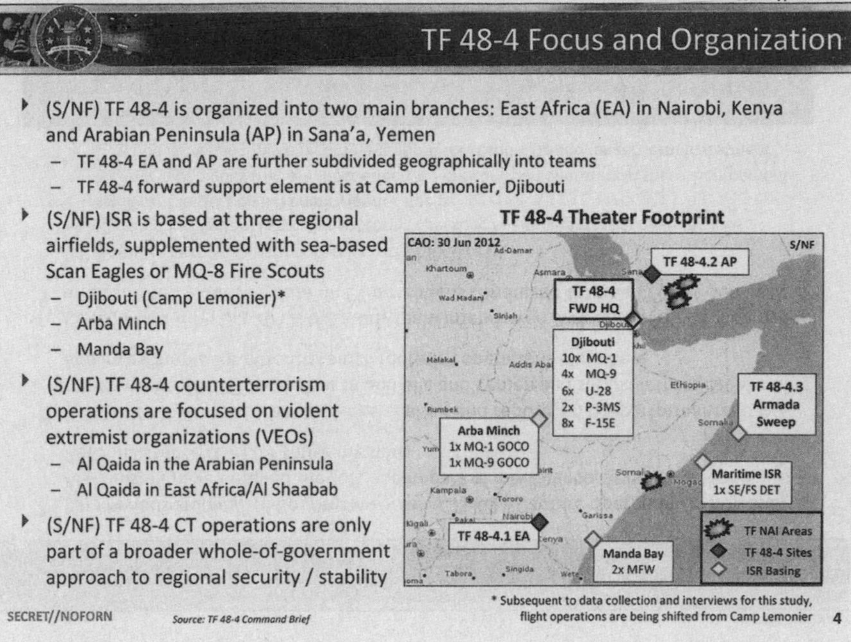

A slide from a classified Pentagon study outlines the air and naval assets of the secret task force charged with hunting down, killing, and capturing high-value individuals in Somalia and Yemen.

run the drone war in Yemen. At times this drama played out on the pages of the *New York Times* and the *Washington Post*, with leaks coming from both sides in an effort to influence policy. The CIA's backers in Congress argued that the agency showed more "patience and discretion" in its drone strikes, while some prominent military advocates portrayed the agency as ill-equipped to conduct military-style operations and less accountable to Congress.[9]

At the peak of this bureaucratic civil war, in 2012, an influential and well-funded Defense Department entity, the Intelligence, Surveillance, and Reconnaissance Task Force, began preparing a classified brief on challenges faced by special operations forces conducting counterterrorism operations in East Africa and the Arabian Peninsula, such as those operating under TF 48-4.[10] The ISR Task Force fell under the control of Michael Vickers, a powerful veteran of CIA paramilitary operations. Obama had promoted Vickers to undersecretary of defense for intelligence in 2010, and as the

Pentagon's top intelligence official he exerted great influence over matters of covert operations.

The ISR Task Force had been established in 2008 to study the intelligence and surveillance needs of war fighters in Afghanistan and Iraq.[11] By 2012 it had evolved into a multibillion-dollar advocacy wing pushing for the purchase of new surveillance technologies to support the military's black ops forces in waging unconventional wars. The purpose of the ISR study, in outlining the challenges faced by special operations units tasked with disrupting and destroying terrorist networks, was to press for more tools and to collect data that would guide future operations.

Michael G. Vickers, former undersecretary of defense for intelligence, was a powerful figure in the world of covert operations.

On October 15, 2015, *The Intercept* published two versions of the study, both titled "ISR Support to Small Footprint CT [Counterterrorism] Operations – Somalia/Yemen." One slide deck, a detailed report, was distributed in February 2013, and the other, an executive summary, was circulated in May 2013, the month Obama gave the first major address of his presidency on drones and targeted killings. The timing of the reports is interesting, because it was during this period that the Obama administration began to publicly advance the idea of handing control of the drone program to the military.

"The United States has taken lethal, targeted action against al Qaeda and its associated forces, including with remotely piloted aircraft commonly referred to as drones," Obama said in front of a military audience. "As was true in previous armed conflicts, this new technology raises profound questions – about who is targeted, and why; about civilian casualties, and the risk of creating new enemies; about the legality of such strikes under U.S. and international law; about accountability and morality."[12] Drone strikes, he asserted, are "effective" and legal. Obama did not mention TF 48-4 in his speech, but it was the actions of that special operations task force – and those of the CIA's parallel program – that he was discussing.

During the period covered in the ISR study, January 2011 through

On May 23, 2013, President Obama gave his first formal address on drone strikes, at the National Defense University in Washington.

June 2012, three U.S. citizens were killed in drone strikes in Yemen. Only one, the radical preacher Anwar al Awlaki, was labeled the intended target of a strike. The United States claimed it did not intend to kill Samir Khan, who was traveling with Awlaki when a Hellfire missile hit their vehicle. The third – and most controversial – killing of a U.S. citizen was that of Awlaki's son, sixteen-year-old Abdulrahman Awlaki. He was killed two weeks after his father, while having dinner with his cousin and some friends. Immediately after the strike anonymous U.S. officials asserted that the younger Awlaki was connected to al Qaeda and was in fact twenty-one. After the family produced his birth certificate, the United States changed its position, with an anonymous official calling the killing of the teenager an "outrageous mistake."[13]

A former senior official in the Obama administration who worked on the high-value targeting program and asked not to be identified because he was discussing classified material, told me in 2013 that after the Abdulrahman strike, the president was "surprised and upset and wanted an explanation." "We had no idea the kid was there," the official said. The White House did not officially acknowledge the strikes until nearly two years later. "We killed three U.S. citizens in a very short period," the official continued. "Two of them weren't even targets: Samir Khan and Abdulrahman Awlaki. That doesn't look good. It's embarrassing." The former official said that John Brennan, President Obama's top counterterrorism adviser, "suspected that the kid had been killed intentionally and ordered a review. I don't know what happened with the review."[14] When asked about the review, a spokesperson for the National Security Council told me, "We cannot discuss the sensitive details of specific operations."

Lt. Gen. Flynn, who since leaving the Defense Intelligence Agency has become an outspoken critic of the Obama administration, charges that the White House relies heavily on drone strikes for reasons of expediency rather than effectiveness. "We've tended to say, 'Drop another bomb via a drone and put out a headline that

A still from a home video of Abdulrahman Awlaki playing with his younger siblings in the family's courtyard in 2009. The sixteen-year-old U.S. citizen was killed in a drone strike on October 14, 2011, in Yemen.

we killed Abu Bag of Doughnuts,' and it makes us all feel good for twenty-four hours," Flynn said. "And you know what? It doesn't matter. It just made them a martyr. It just created a new reason to fight us even harder."

But Glenn Carle, a former senior CIA officer, disputes Flynn's characterization of the Obama administration's motive in its widespread use of drones. Carle, who spent more than two decades in the CIA's clandestine services, told me, "I would be skeptical the government would ever make that formal decision to act that way. Obama is always attacked by the right as being soft on defense and not able to make the tough decisions. That's all garbage. The Obama administration has been quite ruthless in its pursuit of terrorists. If there are people who we, in our best efforts, assess to be trying to kill us, we can make their life as short as possible. And we do it."

According to the ISR study, TF 48-4 did in fact have an impressive cache of firepower in Djibouti to kill or capture people approved for the kill list by the president. According to one slide, as of 2012 the base at Camp Lemonnier housed more than a dozen armed drones and additional surveillance aircraft. Its arsenal also included eight manned F-15E warplanes, which can carry so-called bunker busters – 5,000-pound laser-guided bombs. The ISR Task Force recommended

Lt. Gen. Michael Flynn, former head of the Defense Intelligence Agency, was an architect of JSOC's "find, fix, finish" doctrine.

providing special operations forces with more and better drones and an enhanced mandate to capture and interrogate suspects "via host-nation partners."

Outsourcing U.S. kill/capture operations to local forces, which occurred frequently during the Bush administration, regularly led to human rights abuses, torture, and extrajudicial killings. "I'm very hesitant on backing foreign militaries or paramilitary forces or militias," said Clinton Watts, a former FBI special agent who worked on counterterrorism and later served as an executive officer of the Combating Terrorism Center at West Point. "I've seen that up close before, and you're backing rape and pillage campaigns through the countryside usually. You can't control them, and you don't have transparency over what they do, and it blows up in your face the same way that a bad drone strike does."

The ISR Task Force asserted that an increase in the number of capture operations could be achieved by using U.S. "advisors" to build partnerships with local forces and by conducting "advance force operations." AFOs are used by the U.S. military to discreetly plant tracking devices, conduct surveillance, and physically access places inhabited by potential targets, often in "denied areas" where the United States is not yet at war. Forces deployed in AFOs may also conduct clandestine "direct actions," including kill/capture operations.[15] During the Bush administration AFOs served as a primary vehicle for justifying the clandestine deployment of U.S. special operators across the world to engage in "operational preparation" of a future battlespace.[16] Those activities expanded after 9/11 as the Bush administration adopted the view that "the entire world is the 'battlespace.'"[17]

A July 2015 U.S. government contract solicitation for training Pacific Command personnel who conduct AFOs envisions a course focusing on tactics "that directly or indirectly support technical surveillance operations in non-permissive environments."[18] Among them, breaking and bypassing a slew of locks, both physical and digital; cloning hotel room key cards; picking advanced car lock systems; and learning "physical restraint escape techniques." The solicitation

PHASE I:
IDENTIFICATION AND FUNCTION OF VARIOUS LOCK FAMILIES
WARDED, BIT KEY, DISC TUMBLER AND PIN TUMBLER LOCKS
WHEEL AND DIAL COMBINATION LOCKS
PADLOCK BYPASS TECHNIQUES
FOREIGN PADLOCK DEFEAT EXERCISE
SIMPLEX AND MECHANICAL PUSH BUTTON LOCKS
SCHLAGE WAFER LOCK PICKING AND DECODING
KEY CASTING AND KEY IMPRESSIONING
BUILDING ENTRY OPERATIONS AND DOOR HARDWARE BYPASS TECHNIQUES
TUBULAR KEY AND CRUCIFORM KEY LOCKS
MASTER KEY SYSTEM EXPLOITATION
PIN TUMBLER DECODING AND SIGHT READING
DIGITAL DOOR HARDWARE AND CLICK LOCKS
KEY MENSURATION AND USE OF CODE MACHINES AND KEY DUPLICATORS
WORKING WITH PIN TUMBLER CYLINDERS - MORTISE, RIM, EUROPROFILE, KEY-IN-KNOB
HOTEL MAG-STRIPE CARD KEY CLONING
RFID CREDENTIAL CLONING
WORKING WITH EUROPROFILE CYLINDERS
PICKING AND DECODING ROTATING DISC TUMBLER LOCKS
PICKING AND IMPRESSIONING DIMPLY KEY LOCKS
ADVANCED KEY IMPRESSIONING TECHNIQUES

PHASE II:
VISUAL DECODING OF EDGE CUT AND HIGH SECURITY AUTOMOBILE KEYS
PICKING, DECODING, GENERATING KEYS TO EUROPEAN FORD / JAGUAR TIBBE LOCKS
GENERATING KEYS TO AUTOMOTIVE DIMPLE KEY LOCKS
VISUAL TUMBLER DECODING TECHNIQUES
ADVANCED AUTOMOTIVE KEY IMPRESSIONING
USE OF VARIOUS MECHANICAL PICKING AND DECODING TOOLS
AUTOMOTIVE KEY GENERATION - STANDARD AND HIGH SECURITY VEHICLE LOCKS
USE OF HAND HELD CLIPPERS TO GENERATE HIGH SECURITY KEYS BY CODE
USE OF ELECTRONIC CODE MACHINES TO GENERATE AND DUPLICATE HIGH SECURITY KEYS
SELECTED TRANSPONDER BYPASS TECHNIQUES
VEHICLE ACCESS AND KEY GENERATION PRACTICAL EXERCISES

PHASE III:
PHYSICAL RESTRAINT ESCAPE TECHNIQUES, TOOL CONSTRUCTION AND CONCEALMENT
MITIGATION OF OVERSEAS OPERATIONAL CONSTRAINTS
TACTICAL PLANNING CONSIDERATIONS
CTR IN SUPPORT OF BUILDING ENTRY AND KEY GENERATION OPERATIONS
CTR IN SUPPORT OF VEHICLE ACCESS OPERATIONS
OPEN SOURCE TOOLS AND DATA BASES IN SUPPORT OF CTR OPERATIONS
RESTRICTED DATA BASES IN SUPPORT OF CTR OPERATIONS
USE OF TECHNICAL COLLECTION TOOLS IN SUPPORT OF KEY GENERATION MISSIONS
CULMINATING TEAM TACTICAL EXERCISES

U.S. military descriptions of the “skill set” required for advance force operations.

stated that operatives need such courses to "remain proficient in this highly refined skill set."

The tone of the ISR study at times gives the impression that special operations forces were effectively prisoners of resource shortages and a legal bureaucracy that interfered with the military's ability to kill or capture terrorists with the frequency, efficiency, and urgency demanded by policymakers. Those sentiments were echoed by Lt. Gen. Flynn, who served for years as the chief intelligence officer for JSOC. "You cannot conduct counterinsurgency, counterterrorism, or counterguerrilla operations without having effective interrogation operations," Flynn said in an interview. "If the president says, 'Defeat this enemy,' but you say you need resources that you never get, you just can't defeat the enemy. Without the ability to capture or interrogate, your effectiveness when conducting counterterrorism operations can be cut in half, if not even lower than that, and that's the challenge that we face."

Carle, the former CIA officer, said the ISR study is part of the "classic" turf war: "If you get the budget, then you control the decisions, and everybody thinks that whatever toys they control are the toys that need to be used and therefore you need more of them." The Pentagon wants "to expand their influence," he added, "because then you don't have obstreperous and disheveled civilian CIA guys who clink glasses in salons telling you how to do things. They don't want that. That's a classic turf institutional tension."

The ISR study, which utilizes corporate language to describe lethal operations as though they were a product in need of refining and upgrading, includes analyses from IBM, which has boasted that its work for the Pentagon "integrates commercial consulting methods with tacit knowledge of the mission, delivering work products and advice that improve operations and creates [sic] new capabilities."[19] The study compares the tempo and methods of conventional operations in which U.S. personnel were on the ground in large numbers, as in Iraq and Afghanistan, to the shadow wars in Yemen and Somalia, where there was a scant and sporadic U.S. military presence. Unlike in Iraq and Afghanistan, where special operations units were given carte blanche to engage in a systematic kill/capture program, in Somalia and Yemen they were required to operate under more stringent rules and guidelines. "When compared to previous operations," the study asserts, "the amount of time required to action objectives is literally orders of magnitude higher." The CIA has operated in Pakistan with looser requirements for obtaining the president's direct approval before launching strikes; the president also waived the requirement that a CIA target present an "imminent" threat.[20]

"Relatively few high-level terrorists meet criteria for targeting under the provisions," the study states, and requiring "low" collateral damage and adherence to the "near certainty" standard for positively identifying a target "reduces targeting opportunities." The authors of the study lament the technical difficulties of achieving positive identification of a targeted person and guaranteeing minimal collateral damage, particularly when insufficient numbers of drones and full-motion video platforms cause "blinking" in the surveillance apparatus. One former senior special operations officer, who asked not to be identified because he was discussing classified material, said that the ISR study was best understood as a "bitch brief." The message, he said, was this: "We can't do what you're asking us to do because you are not giving us the resources to get it done."

As the Obama era draws to a close, the internal debate over control of the drone program continues, with some reports suggesting the establishment of a "dual command" structure for the CIA and the military.[21] For now it seems that the military is getting much of what it agitated for in the ISR study. In August 2015 the Wall Street Journal reported that the military plans to "sharply expand the number of U.S. drone flights over the next four years, giving military commanders access to more intelligence and greater firepower to keep up with a sprouting number of global hot spots." The paper claimed that drone flights would increase by 50 percent by 2019, adding, "While expanding surveillance, the Pentagon plan also grows the capacity for lethal airstrikes."[22]

39←

EKIA

→83

Over a period of four and a half months in 2012, U.S. forces used drones and other aircraft to kill 155 people in northeastern Afghanistan. Nineteen were jackpots; the remaining, 136 people, were classified as EKIA.

HAYMAKER Operations (01 May – 15 Sep 2012)					
Type	# Ops	EKIA	Detainees	JP	%
Enabled Ops	27	2	61	13	48%
Kinetic Strikes	27	155	N/A	19	70%
Total	54	157	61	32	

Note the "%" column. It specifies the number of jackpots (JPs) divided by the number of operations: a 70 percent success rate. But it ignores well over a hundred other people killed along the way.

This means that almost nine out of ten people killed in these strikes were not the intended targets.

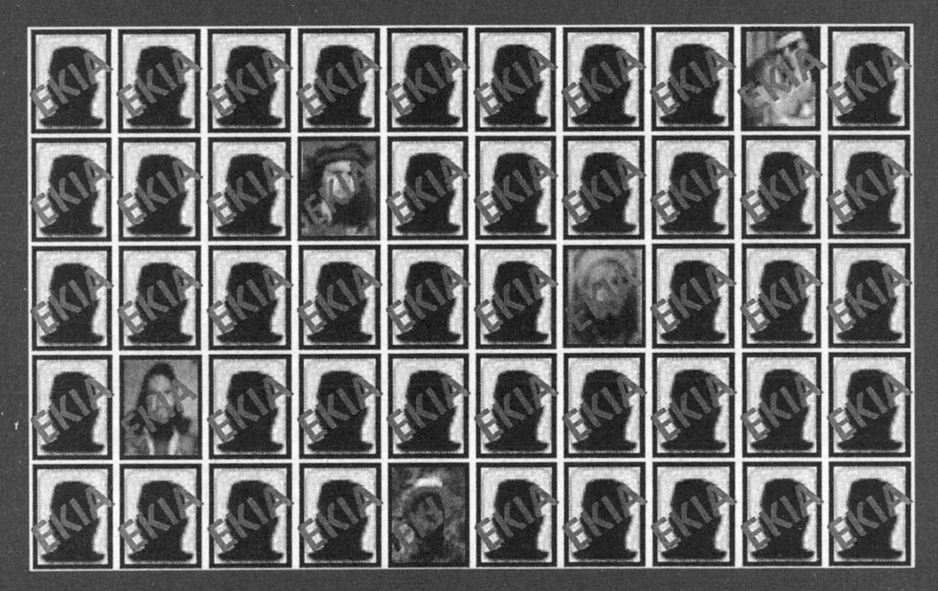

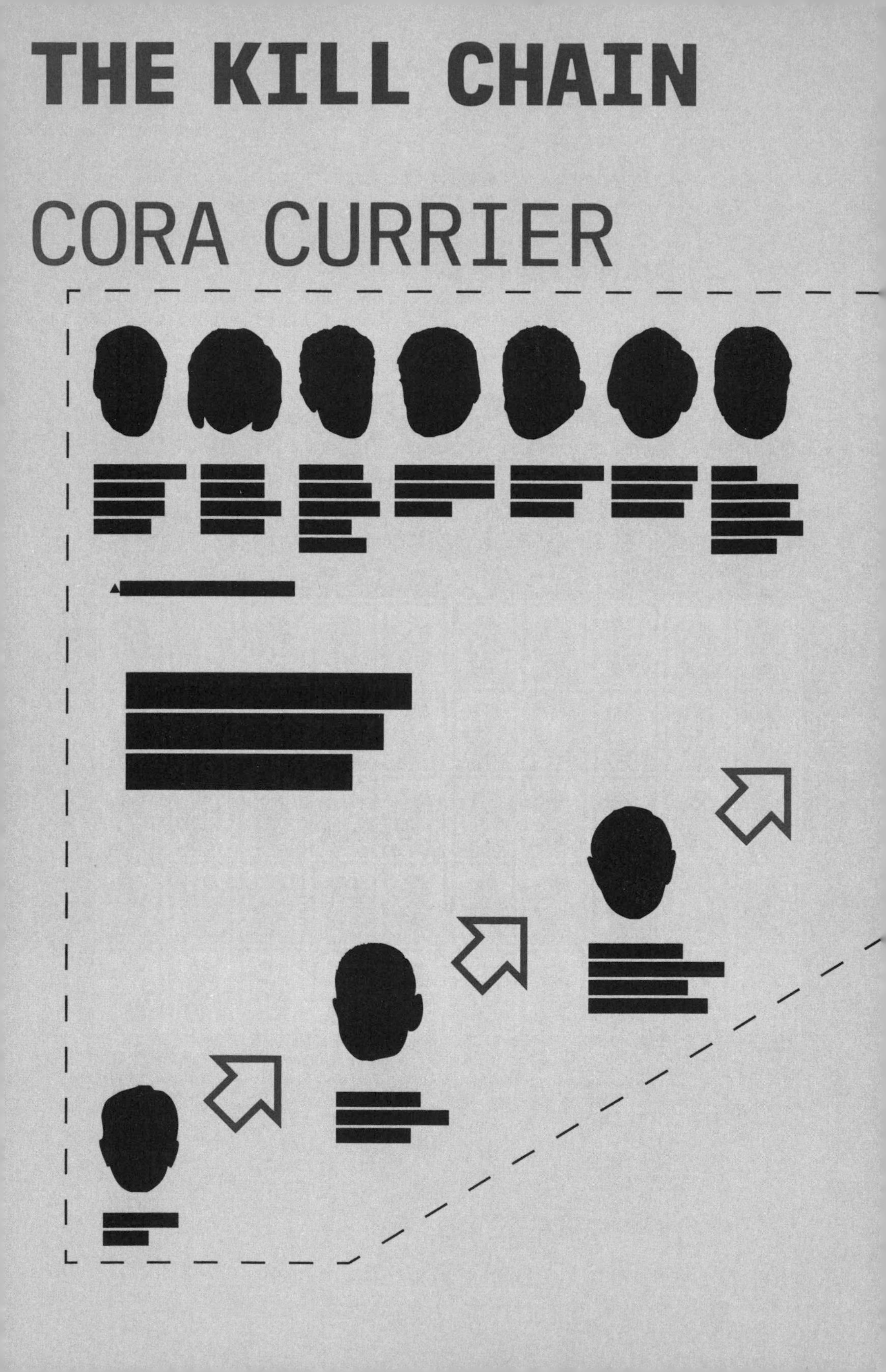
THE KILL CHAIN
CORA CURRIER

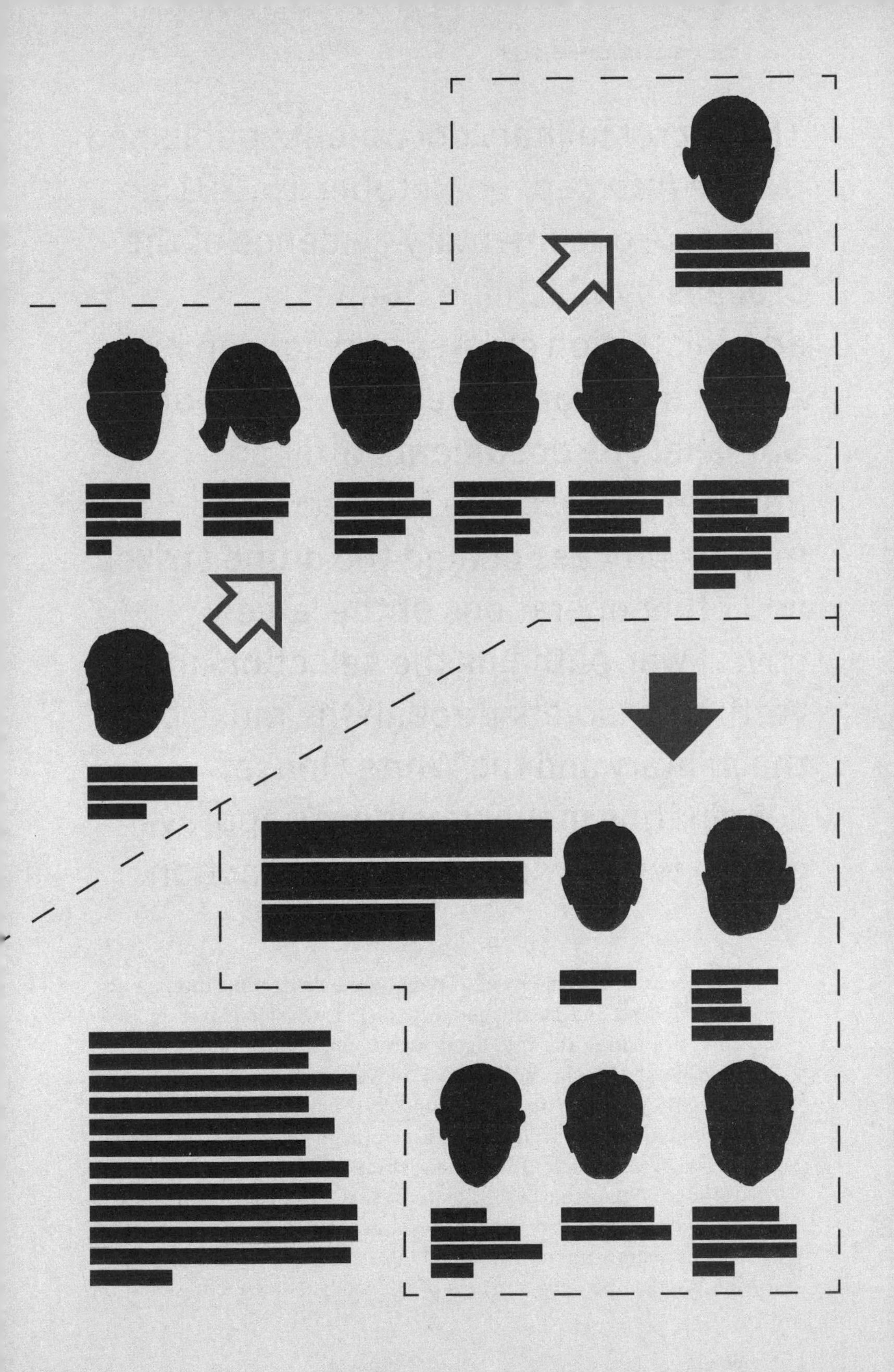

The secret military documents published by *The Intercept* on October 15, 2015, offer rare documentary evidence of the process by which the Obama administration creates and acts on its kill list of terror suspects in Yemen and Somalia. The documents offer an unusual glimpse into the decision-making process behind the drone strikes and other operations of the largely covert war, outlining the selection and vetting of targets through the ranks of the military and the White House, culminating in the president's approval of a sixty-day window for lethal action.

The documents come from a study by the Pentagon's Intelligence, Surveillance, and Reconnaissance Task Force, circulated in early 2013, evaluating the intelligence and surveillance technology behind JSOC's killing campaign in Yemen and Somalia in 2011 and 2012. They illuminate and in some cases contradict the administration's public description of a campaign directed at high-level terrorists who pose an imminent threat to the United States. The authors of the study admit frankly that capturing terrorists is a rare occurrence, and they hint at the use of so-called signature strikes against unknown individuals exhibiting suspicious behavior. We obtained two versions of the study, a longer presentation

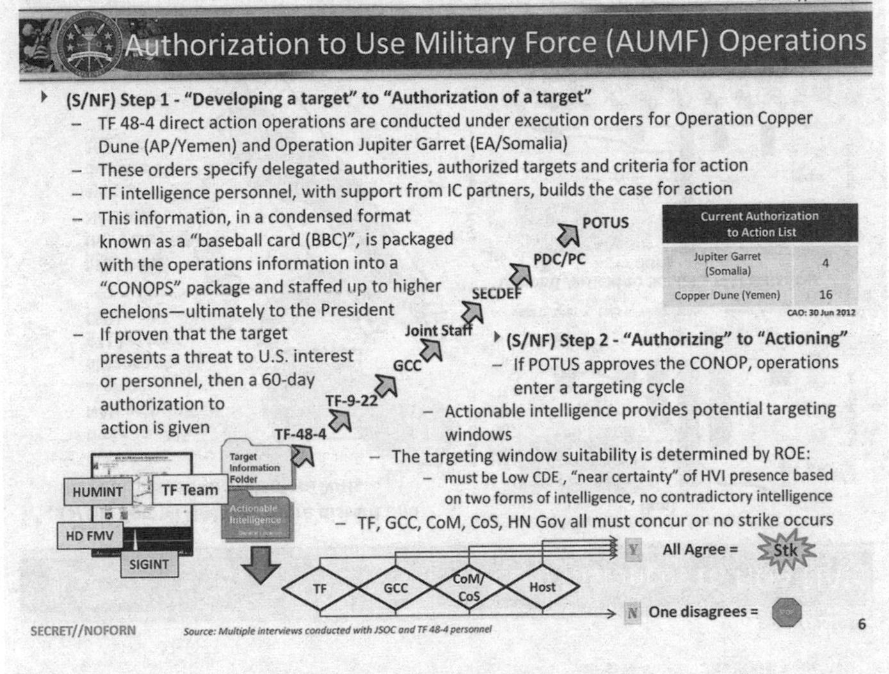

A slide from a May 2013 Pentagon presentation shows the chain of command for ordering drone strikes and other operations carried out by JSOC in Yemen and Somalia.

dated February 2013 and an executive summary from May 2013, which includes a slide showing the chain of command leading to the approval of a lethal strike.

The Obama administration has been loath to declassify even the legal rationale for drone strikes, let alone detail the bureaucratic structure revealed in these documents. Both the CIA and JSOC conduct drone strikes in Yemen, and very little has been officially disclosed about either operation. Civil rights advocates welcomed the disclosure. "The public has a right to know who's making these decisions, who decides who is a legitimate target, and on what basis that decision is made," said Jameel Jaffer, deputy legal director of the American Civil Liberties Union. Both the Pentagon and the National Security Council declined to respond to detailed questions about the study or about the drone program more generally. The NSC would not say if the process for approving targets or strikes had changed since the study was produced.

THE CHAIN OF COMMAND

Tom Donilon National Security Adviser

Hillary Clinton Secretary of State

Timothy Geithner Secretary of the Treasury

Leon Panetta Secretary of Defense

Eric Holder Attorney General

Steven Chu Secretary of Energy

Janet Napolitano Secretary of Homeland Security

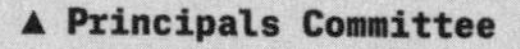

▲ Principals Committee

STEP ONE: CHOOSING A TARGET

Gen. Martin Dempsey Chairman of the Joint Chiefs of Staff

Gen. James Mattis Centcom Commander

JSOC Task Force

Barack Obama
President of the United States

Jeffrey Zients
Director of OMB

Susan Rice
Ambassador to the UN

Jacob Lew
White House Chief of Staff

John Brennan
Counter-terrorism Adviser

James Clapper
Director of National Intelligence

Gen. Martin Dempsey
Chairman of the Joint Chiefs of Staff

Leon Panetta
Secretary of Defense

STEP TWO: TAKING A STRIKE

JSOC Task Force

Gen. James Mattis
Centcom Commander

Gerald Feierstein
Ambassador to Yemen

CIA Station
Chief in Yemen

Abdu Rabbu Mansour Hadi
President of Yemen

According to a Pentagon study obtained by *The Intercept*, President Obama signed off on 60-day authorizations to kill suspected terrorists, but did not sign off on individual strikes. This graphic shows the officials who would have been involved in approving targets in Yemen in early 2012, according to the chain of command laid out in the study.

TWO STEPS TO A KILL

The May 2013 slide describes a two-part process of approval for an attack: (1) "'Developing a target' to 'Authorization of a target'"; (2) "'Authorizing' to 'Actioning.'" According to the slide, intelligence personnel from JSOC's Task Force 48-4, working alongside other intelligence agencies, would build the case for action against an individual, eventually generating a "baseball card" on the target, which was "staffed up to higher echelons – ultimately to the president."

The intelligence package on the person being targeted passed from the JSOC task force tracking him to the command in charge of the region – U.S. Central Command for Yemen and U.S. Africa Command for Somalia – and then to the Joint Chiefs of Staff, followed by the secretary of defense. It was then examined by a circle of top advisers known as the Principals Committee of the National Security Council, and their seconds in command, known collectively as the Deputies Committee. The slide detailing the kill chain indicates that while Obama approved each target, he did not approve each individual strike, although news accounts have previously reported that the president personally "signs off" on strikes outside of Afghanistan or Pakistan.[1] However, the slide does appear to be consistent with Obama's comment in 2012 that "ultimately I'm responsible for the process."[2]

There have been various accounts of this drone bureaucracy, and almost all stress the role of Obama's influential counterterrorism adviser John Brennan, who became director of the CIA in 2013, and of top administration lawyers in deciding who could be killed. Under Brennan the nominations process was reportedly concentrated in the White House, replacing video conferences once run by the Joint Chiefs of Staff and elevating the role of the National Counterterrorism Center in organizing intelligence.[3] Later in 2013 the White House reportedly tightened control over individual strikes in Yemen.[4]

At the time of the ISR study, with the president's approval, JSOC had a sixty-day window to hit a target. For the actual strike the task force needed approval from the Geographic Combatant Command as well as the ambassador and CIA station chief in the country where the target was located. For a very important target, such as Anwar al Awlaki, a U.S. citizen linked to al Qaeda, "it would take a high-level official to approve the strike," said Lt. Col. Mark McCurley, a former drone pilot who worked on operations in Yemen and published a book about his experiences. "And that includes a lot of lawyers and a lot of review at different levels to reach that decision. We have an extensive chain of command, humans along the whole link that monitor the entire process from start to finish on an airstrike."[5] The country's government was also supposed to sign off. "One Disagrees = STOP," the slide notes, accompanied by a tiny red stop sign.

In practice, however, the degree of cooperation with the host nation has varied. Somalia's minister of national security, Abdirizak Omar Mohamed, said in an interview that the United States alerted Somalia's president and foreign minister of strikes "sometimes ahead of time, sometimes during the operation. . . . normally we get advance notice." He said he was unaware of an instance when Somali officials had objected to a strike, but added that if they did, he assumed the United States would respect Somalia's sovereignty.

By 2011, when the study's time frame began, Yemen's president Ali Abdullah Saleh was in crisis. Facing domestic protests during the Arab Spring, he left the country in June 2011 after being injured in a bombing. Both the CIA and JSOC stepped up their drone campaigns, which enjoyed vocal support from Saleh's eventual successor, Abdu Rabbu Mansour Hadi.[6] "It was almost never coordinated with Saleh. Once Hadi became president, March 2012, there was a big chance we'd be in the loop" before drone strikes were conducted, explained a former senior Yemeni official who worked for both the Saleh and the Hadi government. In 2015, with Yemen's capital under the control of the Houthi rebel group and undergoing bombardment by Saudi Arabia, administration lawyers did not seem worried about asking permission to carry out drone strikes amid the fray.[7] "Now I think they don't even bother telling anyone. There is really no one in charge to tell," said the former Yemeni official, who requested anonymity, citing current unrest and the fact that he no longer works for the government.

WHO CAN BE TARGETED

Both the Bush and the Obama administrations have maintained that the 2001 Authorization for Use of Military Force permits the pursuit of members of al Qaeda and its affiliates wherever they may be located. The ISR study refers throughout to operations that fall under AUMF. But it also underlines how the targeted killing campaigns differ from traditional battlefields, noting that the region is located "outside a defined theater of active armed conflict," which limits "allowable U.S. activities."

Obama administration officials have said that in addition to being a member of al Qaeda or an associated force, targets must also pose a significant threat to the United States.[8] In May 2013, facing increasing pressure to fully admit the existence of the drone war and especially to address allegations of civilian harm, the White House released policy guidelines for lethal counterterrorism operations that seemed to further restrict such actions.[9] Obama announced that action would be taken only against people who posed a "continuing, imminent threat to the American people" and who could not be captured.[10] A strike would occur only with "near certainty" that no civilians would be killed or injured.

Even with the new guidelines, legal observers, particularly human rights lawyers, have disputed the Obama administration's position that the United States, in strict legal terms, is in an armed conflict with al Qaeda in Yemen or Somalia; they therefore dispute what standards should apply to strikes.[11] Others question the extent to which the hundreds of people killed in drone strikes in those countries meet the supposedly strict criteria. "I think there can be questions raised about how stringently some of the requirements are being applied," said Jennifer Daskal, an assistant professor of law at American University who worked for the Department of Justice from 2009 to 2011. "Near certainty of no civilian deaths: is that really imposed? What does it mean for capture not to be feasible? How hard do you have to try?"

It is not clear whether the study reflects the May policy guidance, since it does not give an extensive description of the criteria for approving a target, noting only that the target must be "a threat to U.S. interest or personnel." A spokesperson for the National Security Council would not explain why the standards in the study differ from the guidelines laid out in May 2013 but emphasized that "those guidelines remain in effect today."

The two-month window for striking, said Hina Shamsi, director of the ACLU's National Security Project, shows the administration's broad interpretation of "a continuing, imminent threat": "If

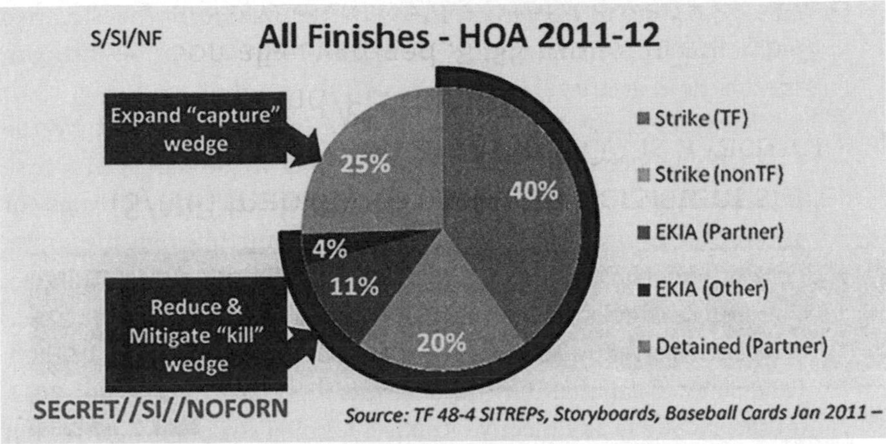

The ISR study recommended more captures, rather than killings, because of the intelligence that could be gleaned from interrogations and collected materials.

you have approval over a months-long period, that sends the signal of a presumption that someone is always targetable, regardless of whether they are actually participating in hostilities."

The slide illustrating the chain of approval makes no mention of evaluating options for capture. It may be implied that those discussions are part of the target development process, but the omission reflects the brute facts beneath the Obama administration's stated preference for capture: detention of marked targets is incredibly rare.

A chart in the study shows that in 2011 and 2012 captures accounted for only 25 percent of operations carried out in the Horn of Africa, and all were apparently by foreign forces. In one of the few publicized captures of the Obama presidency, al Shabaab commander Ahmed Abdulkadir Warsame was picked up in April 2011 by U.S. forces in the Gulf of Aden and brought to Manhattan for trial, though he may not be reflected in the study's figures because he was apprehended at sea.

The study does not contain an overall count of strikes or deaths, but it does state that "relatively few high-level terrorists meet criteria for targeting" and that at the end of June 2012 there were sixteen authorized targets in Yemen and only four in Somalia. Despite the small number of people on the kill list, in 2011 and 2012 there were at least fifty-four U.S. drone strikes and other attacks reported in Yemen, killing a minimum of 293 people, including fifty-five civilians, according to figures compiled by the Bureau of Investigative Journalism. In Somalia there were at least three attacks, resulting in the deaths of at least six people.[12]

Some of those Yemen strikes were likely carried out by the CIA, which since mid-2011 has flown drones to Yemen from a base in Saudi Arabia and reportedly has its own kill list and rules for strikes.[13] It is also clear that the military sometimes harmed multiple other people in trying to kill one of those high-level targets. The study includes a description of the hunt for an alleged al Qaeda member referred to as "Objective Rhodes" or "Anjaf," who is likely Fahd Saleh al-Anjaf al-Harithi, who was reported killed in July 2012, on the same day as Objective Rhodes. A failed strike on Harithi that April killed two "enemies." News accounts at the time reported that three "militants" had died.[14]

The large number of reported strikes may also be a reflection of signature strikes in Yemen, where people can be targeted based on patterns of suspect behavior. In 2012 administration officials said that President Obama had approved strikes in Yemen on unknown people, calling them TADS, or "terror attack disruption strikes," and claiming that they were more constrained than the CIA's signature strikes in Pakistan.[15] The study refers to using drones and spy planes to "conduct TADS related network development," presumably a reference to surveilling behavior patterns and relationships in order to carry out signature strikes. It is unclear what authorities govern such strikes, which undermines the administration's insistence that the United States kills mainly "high-value" targets.

NEAR CERTAINTY

According to the White House guidelines released in May 2013, the decision to strike should be based on thorough surveillance and occur only in the absence of civilians. A strike requires "near certainty that the terrorist target is present" and "near certainty that non-combatants will not be injured or killed." The ISR study describes the rules for a strike slightly differently, stating that there must be a "low CDE [collateral damage environment]," meaning a low estimate of how many innocent people might be harmed. It also states there must be "near certainty" that the target is present, "based on two forms of intelligence" with "no contradictory intelligence." In

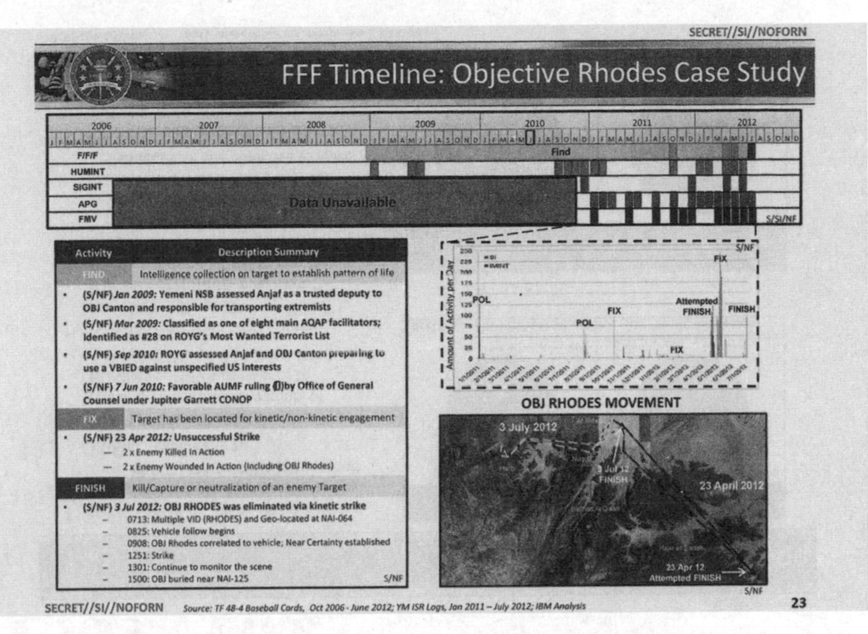

A slide from the ISR study recounts the hunt for an alleged al Qaeda member (likely Fahd Saleh al-Anjaf al-Harithi), showing that two others died in a botched attempt to kill him.

contrast to the White House statement, the "near certainty" standard is not applied to civilians. Although the study cites the "need to avoiding [*sic*] collateral damage areas" as a reason for "unsuccessful" missions, it does not give numbers of civilian casualties or examples of bad intelligence leading to a mistaken kill.

Yet the overall conclusion is that getting accurate positive identification is a "critical" issue for the drone program in the region, due to limitations in technology and the number of spy aircraft available. The military relies heavily on signals intelligence – drawn from electronic communications – and much of it comes from foreign governments, who may have their own agendas. Of course, identifying the correct target relates directly to the issue of civilian casualties: if you don't have certainty about your target, it follows that you may well be killing innocent people. In Iraq and Afghanistan, "when collateral damage did occur, 70 percent of the time it was attributable to failed – that is, mistaken – identification," according to a paper by Gregory McNeal, an expert on drones and security at Pepperdine School of Law.[16] Another factor is timing: a former senior special operations officer, who asked not to be identified because he was discussing classified material, explained that if the sixty-day strike

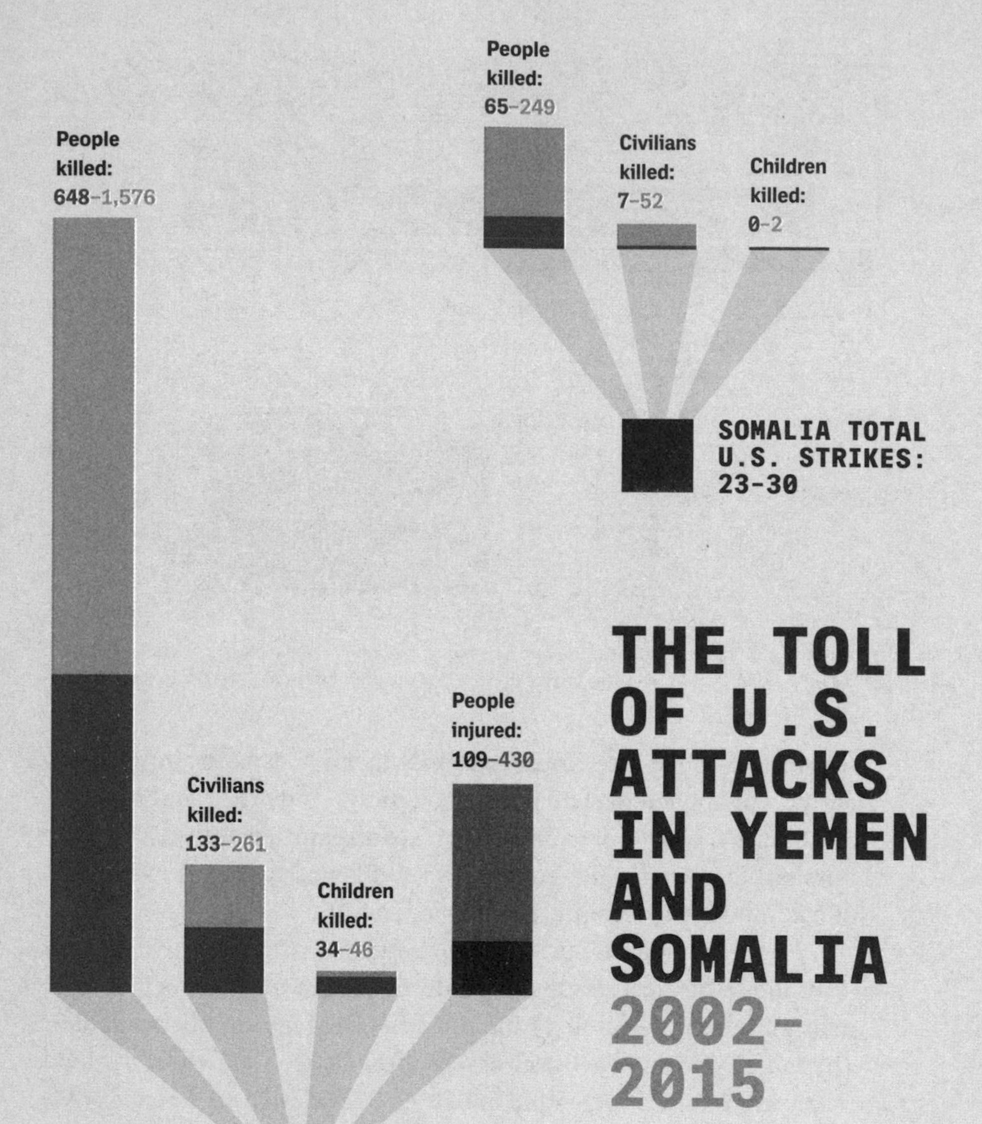

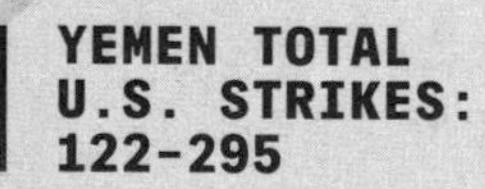

Since the first drone strike in Yemen in 2002, hundreds of people have been killed in U.S. operations in Yemen and Somalia, many of them innocent civilians. The tallies shown here were compiled by the Bureau of Investigative Journalism from reports of both CIA and JSOC drone strikes and other operations. The large range in the estimates is due to the inherent difficulties of collecting data on airstrikes in war zones. The identities of the "people killed" were often unknown and may include civilians as well as suspected terrorists or militants. The United States almost never publicly acknowledges individual operations.

authorization expired, analysts would have to start over in building the intelligence case against the target. That could lead to pressure to take a shot while the window was open.

During the time of the ISR study there were multiple well-reported, high-profile incidents in which JSOC strikes killed the wrong people. Perhaps most famous is the case of Abdulrahman Awlaki, the sixteen-year-old son of Anwar al Awlaki, who died in October 2011 in a JSOC strike while eating dinner with his cousins, two weeks after his father was killed by a CIA drone. In press accounts one anonymous official called Abdulrahman's death "an outrageous mistake," although others said he was with people believed to be members of al Qaeda in the Arabian Peninsula.[17] Publicly the government has said only that he "was not specifically targeted." A September 2012 strike in Yemen, extensively investigated by Human Rights Watch and the Open Society Foundations, killed twelve civilians, including three children and a pregnant woman.[18] No alleged militants died in the strike, and the Yemeni government paid restitution for it, but the United States never offered an explanation. "The mothers and fathers and brothers and sisters of the people who were killed in these drones strikes want to know why," said Amrit Singh, senior legal officer at the Open Society Justice Initiative. "We're left with no explanation as to why they were targeted and in most cases no compensation, and the families are aware of no investigation."

In spring 2015, in a rare admission of a mistake in targeting, the White House announced that two hostages held by al Qaeda, an American and an Italian, had been killed in a CIA drone strike in Pakistan in January.[19] Attempting to explain the tragedy, the White House spokesperson used the language of the standards that had failed to prevent it: the hostages had died despite "near certainty," after "near continuous surveillance," that they were not present.

THE HEART OF THE DRONE MAZE
JEREMY SCAHILL

A classified U.S. intelligence document published by *The Intercept* in April 2015 confirms that the sprawling U.S. military base in Ramstein, Germany, serves as the high-tech heart of America's drone program. Ramstein is the site of a satellite relay station that enables drone operators in the American Southwest to communicate with their remote aircraft in Yemen, Somalia, Afghanistan, and other targeted countries. The top-secret slide deck, dated July 2012, provides the most detailed blueprint to date of the technical architecture used to conduct strikes with Predator and Reaper drones.

Amid fierce European criticism of America's targeted killing program, U.S. and German government officials have long downplayed Ramstein's role in lethal U.S. drone operations and have issued carefully phrased evasions when confronted with direct questions about the base. But the slides show that the facilities at Ramstein perform an essential function in lethal drone strikes conducted by the CIA and the U.S. military in the Middle East, Afghanistan, and Africa. The slides were provided by a source with knowledge of the U.S. government's drone program who declined to be identified for fear of retribution. According to the source, Ramstein's importance to the U.S. drone war is difficult to overstate: "Ramstein carries the signal to tell the drone what to do and it returns the display of what the drone sees. Without Ramstein, drones could not function, at least not as they do now."

The new evidence places German chancellor Angela Merkel in an awkward position, given Germany's close diplomatic alliance with the United States. The German government has granted the United States the right to use the property, but only under the condition that the Americans do nothing there that violates German law. The U.S. government maintains that its drone strikes against al Qaeda and its "associated forces" are legal, even outside of declared war zones. But German legal officials have suggested that such operations are justifiable only in actual war zones. Moreover Germany has the right to prosecute "criminal offenses against international law . . . even when the offense was committed abroad and bears no relation to Germany," according to Germany's Code of Crimes against International Law, passed in 2002.[1]

This means that American personnel stationed at Ramstein could, in theory, be vulnerable to German prosecution if they provide drone pilots with data used in attacks. While the German government has been reluctant to pursue such prosecutions, it may come under increasing pressure to do so. "It is simply murder," said Björn Schiffbauer of the Institute for International Law at the University of Cologne. Legal experts interviewed by *Der Spiegel* claimed that U.S. personnel could be charged as war criminals by German prosecutors.[2]

Ramstein is one of the largest U.S. military bases outside the United States, hosting more than sixteen thousand military and civilian personnel. The relay center, which was completed in late 2013, sits in the middle of a massive forest and is adjacent to a baseball diamond used by students at the Ramstein American High School. The large compound, made of reinforced concrete with masonry walls and enclosed in a horseshoe of trees, has a sloped metal roof. Inside this building air force squadrons can coordinate the signals necessary for a variety of drone surveillance and strike missions. On two sides of the building are six massive golf ball-like fixtures known as satellite relay pads.

In a 2010 budget request for the Ramstein satellite station, the U.S. Air Force asserted that without the facility, the drone program

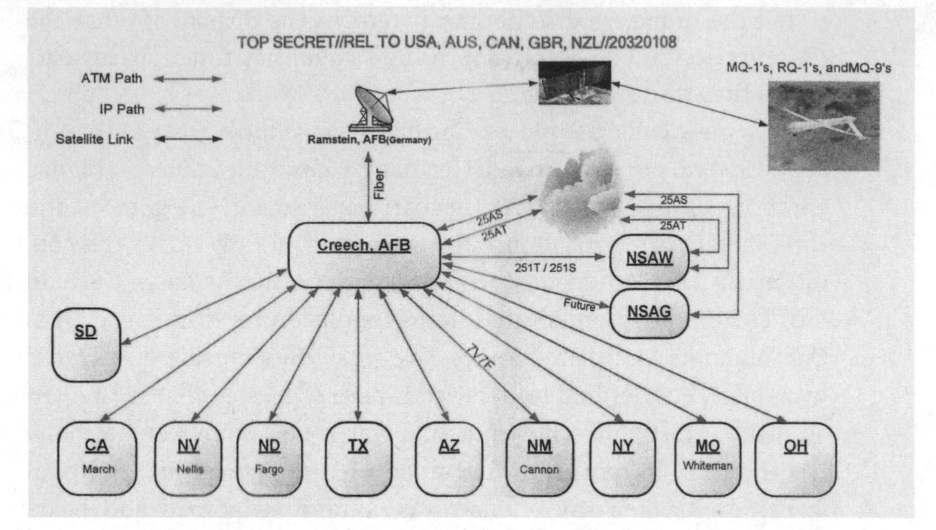

A top-secret slide confirms the central role Germany plays in the U.S. drone war.

could face "significant degradation of operational capability" that could "have a serious impact on ongoing and future missions." Predator and Reaper drones, as well as unmanned Global Hawk aircraft, would "use this site to conduct operations" in Africa and the Middle East, according to the request. It asserted that without the use of Ramstein, drone "weapon strikes cannot be supported."

"Because of multi-theater-wide operations, the respective SATCOM Relay Station must be located at Ramstein Air Base to provide most current information to the war-fighting commander at any time demanded," according to the request. The relay station would also be used to support the operations of a secretive black ops air force program known as "Big Safari." The classified slide deck maps out an intricate spiderweb of facilities across the United States and the globe: from drone command centers on desert military bases in the United States to Ramstein to outposts in Afghanistan, Djibouti, Qatar, and Bahrain and back to NSA facilities in Washington and Georgia. What is clear is that most paths within America's drone maze run through Ramstein.

Creech Air Force Base in Nevada is central to multiple prongs of the U.S. drone war. Personnel stationed at the facility are responsible for drone operations in Afghanistan, which has been on the receiving end of more drone strikes than any country in the world, and Pakistan, where the CIA has conducted a covert air war for the

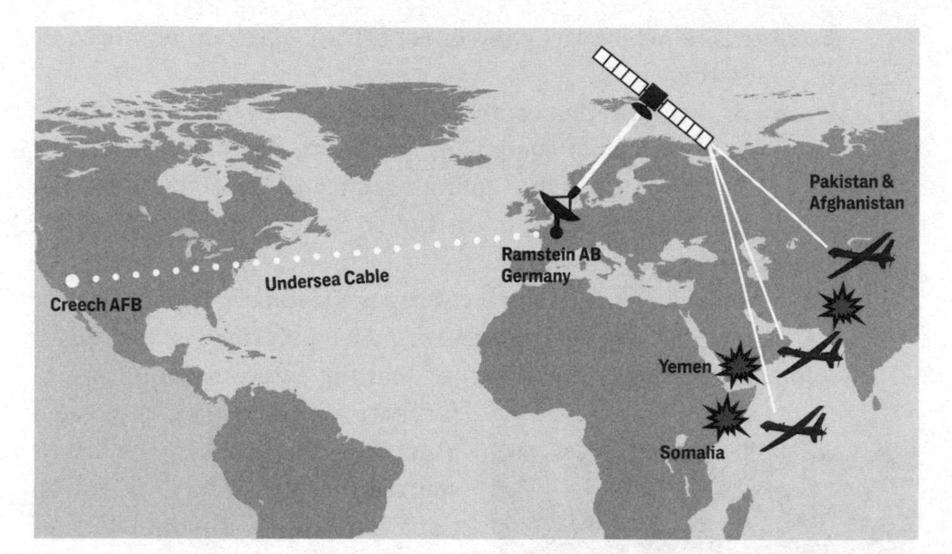

Transatlantic cables connect U.S. drone pilots to their aircraft half a world away.

past decade.[3] The agency's campaign has killed thousands of people, including hundreds of civilians. Some drone missions are operated from other locations, such as Fort Gordon in Georgia and Cannon Air Force Base in New Mexico.

The pilots at Creech and other ground control stations send their commands to the drones via transatlantic fiber optic cables to Germany, where the Ramstein uplink bounces the signal to a satellite that connects to drones over Yemen, Somalia, and other target countries. Ramstein is ideally situated as a satellite relay station to minimize the lag time, or latency between the commands of the pilots and their reception by the aircraft. Too much latency – which would be caused by additional satellite relays – would make swift maneuvers impossible. Video images from a drone could not be delivered to the United States in near real time. Without the speed and precise control an installation like Ramstein allows, pilots would be flying practically blind.

A diagram in the secret document shows how the process works. Ramstein's satellite uplink station is used to route communications between the pilots and aircraft deployed in a variety of countries. Video from the drones is routed back through Ramstein and then relayed to a variety of U.S. intelligence and military facilities around the United States and the globe. Another diagram shows how pilots at Creech connect to Ramstein and then to the Predator Primary

Satellite Link, which facilitates direct control of the drone wherever it is operating.

All of this – location combined with the need to securely house the large quantities of equipment, buildings, and personnel necessary to operate the satellite uplink – has made Ramstein one of the most viable sites available to the United States to serve this critical function in the drone war.

When the German daily newspaper *Süddeutsche Zeitung* and the German public television broadcaster ARD published an exposé on Ramstein in May 2013 and alleged that the base was being used to facilitate drone strikes, it created a massive controversy.[4] The report spurred parliamentary investigations and calls for the United States to explain exactly what it was doing at the base. In response the German and U.S. governments mischaracterized the reporting, and the German government claimed it had no hard evidence of Ramstein's role in lethal strikes. A month later, in a June 2013 speech in Berlin, President Obama addressed the issue of Ramstein's role in the drone war. He did not mention that the satellite relay facility at Ramstein enables U.S. drone strikes. Instead he denied a claim that the journalists had not made: "We do not use Germany as a launching point for unmanned drones . . . as part of our counterterrorism activities."

President Barack Obama and Chancellor Angela Merkel.

In response to questions from *The Intercept*, Pentagon spokesman Maj. James Brindle echoed the precise language of previous government statements. "We maintain robust civilian and military cooperation with Germany and manage all base activities in accordance with the agreements made between the United States and German governments," he said. "The Air and Space Operations Center at Ramstein Air Base conducts operational-level planning, monitoring and assessment of assigned airpower missions throughout Europe and Africa, but does not directly fly or control any manned or remotely piloted aircraft."

The German government has issued similar statements, saying no drone pilots are based at Ramstein and no drones are launched from the base. "The U.S. government has confirmed that such armed and remote aircrafts are not flown or controlled from U.S. bases in Germany," government spokesperson Steffen Seibert said in 2014. In 2013 members of the Bundestag, the German parliament, submitted written questions to their federal government. "To the knowledge of the Federal Government, is it true that U.S. drone attacks in Africa could not be carried out without a special satellite relay station for unmanned flying objects in Ramstein?" the lawmakers asked. "The Federal Government has no reliable information in this regard" was the official reply. Pressed further on the satellite facility and its purpose, the government stated, "The Federal Government has no information regarding the installation of the satellite system or when it started operating."[5]

Internal German government communications provided by *Der Spiegel* show that some German officials tried and failed to get their government to confront the United States about the connection between facilities in Germany and drone strikes. According to a June 2013 document, a senior Foreign Office official, Emily Haber, advocated demanding a clear answer from Washington about the role U.S. facilities in Germany played in drone strikes. Haber was overruled. "The Federal Chancellery and the Defense Ministry would prefer to 'sit out' the pressure from parliament and the public," the response read. The unofficial German-U.S. agreement appears to amount to a "Don't ask, don't tell" understanding.

While most if not all of the official statements by both governments may be technically true, it is also true that without the base, it would be very difficult for the United States to sustain the current drone war. The slide deck uses an array of arrows to show the complex system used to operate drones across the world, and all arrows point to Ramstein. "Everything relies on Ramstein and Creech as central hubs for communication" in both armed and unarmed drone operations, said the source. Aside from the possibility of using an undisclosed satellite uplink station, the only drone operations that would not rely on Ramstein in these regions would be those conducted via aircraft that have a line of sight to a ground control station.

Human rights groups in Germany, as well as opposition politicians, have long suspected that Ramstein has played a direct role in the U.S. drone war. They have called on their government to stop allowing

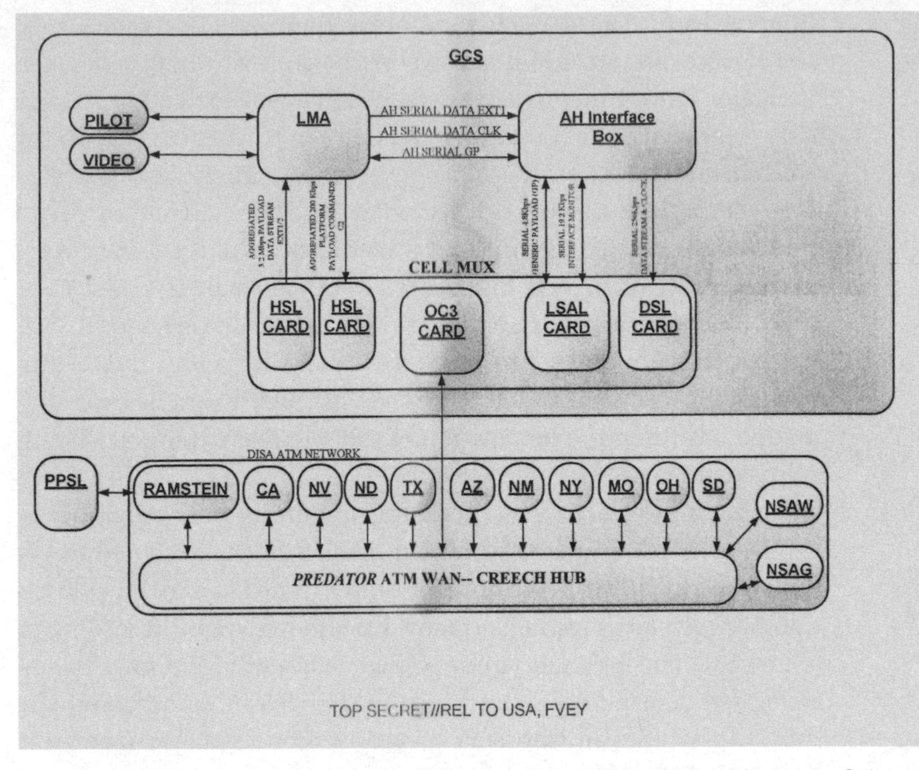

A top-secret slide deck obtained by *The Intercept* shows the complex architecture of the U.S. drone program.

the armed U.S. drone program to operate from German soil. Lt. Gen. David Deptula, the former director of the U.S. Air Force's Combined Air Operations Center, accused such critics of being influenced by "misinformation that's provided by terrorist organizations that these things are being effective against." Deptula oversaw the implementation of the U.S. armed drone program starting in 2001. In an interview he defended the use of drones: "Operations conducted by remotely piloted aircraft really are the most accurate and precise means of applying force. Why would the Germans want to shut down operations that effectively provide information to increase situational awareness of a community of nations that are trying to combat terrorism?"

Kat Craig, the legal director at Reprieve, an international human rights organization that represents victims of drone strikes in Yemen and elsewhere, said the notion that critics of the drone program are being manipulated by propaganda from terrorist organizations

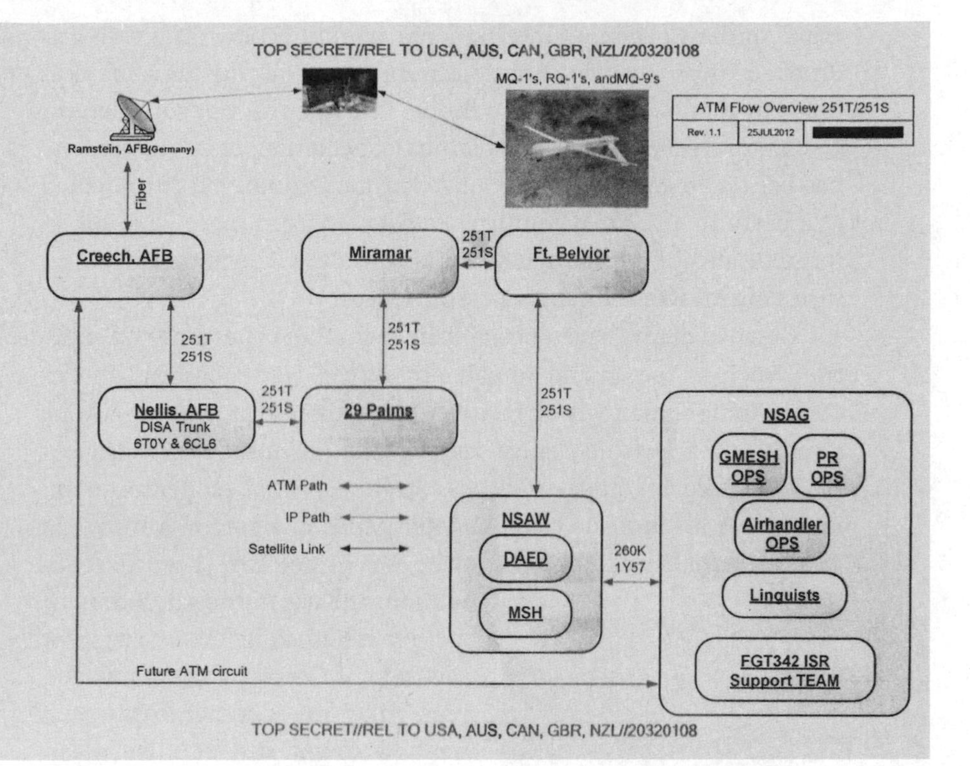

"would be laughable, were it not so offensive towards civilian victims of drone strikes." A report from the Open Society Justice Initiative published in April 2015 found that in nine U.S. drone strikes in Yemen twenty-six civilians were killed, including several children and a pregnant woman.[6] "It has become all too clear," Craig said, "that, too often, those carrying out the strikes simply do not know who they are hitting. This misguided campaign has been allowed free rein because it has been kept hidden from public scrutiny."

While the German government has so far managed to dodge questions on Ramstein's role in drone strikes, the country's judicial system may not have that option. Two related cases have been winding their way through the German legal system. In 2010 a German citizen was killed in a U.S. drone strike in Pakistan. Two years later a federal prosecutor opened a preliminary investigation "to examine whether Bünyamin Erdogan's violent death qualified as a war crime under Germany's international criminal code." The case was dropped after investigators determined that at the time he was killed by a missile fired from a drone, Erdogan was not considered a civilian protected under international law. Rather, they asserted, he had been a "member of an organized, armed group that participated as a party in an armed conflict." Pakistan, according to German interpretations of international law, is considered a war zone in cases involving known militants in certain areas.

German courts haven't established whether other targeted countries, such as Yemen and Somalia, qualify as war zones. In October 2014 a Yemeni man whose relatives were killed in a 2012 U.S. drone strike filed a lawsuit against the German government. Faisal bin Ali Jaber said his brother-in-law, a well-respected moderate imam known for his anti–al Qaeda sermons, and his nephew were killed in a strike. Jaber claimed the strike would not have been possible without the use of the satellite relay facility at Ramstein. "Were it not for the help of Germany and Ramstein, men like my brother-in-law and nephew might still be alive today. It is quite simple: without Germany, U.S. drones would not fly," Jaber said at the time. "I am here to ask that the German people and Parliament be told the full extent of what is happening in their country, and that the German government stop Ramstein being used to help the U.S.'s illegal and devastating drone war in my country." A member of Jaber's legal team accused Germany of "hiding behind status-of-forces agreements," saying the government should "admit its responsibility for civilian deaths caused by U.S. drone warfare."[7]

Yemenis gather around a burned car after it was torched by a drone strike on January 26, 2015. Among the dead was a teenage boy.

On January 20, 2015, the German Defense Ministry submitted a reply to the suit on behalf of the government, which is named as the defendant in the case: "The defendant denies, by claiming ignorance, that the satellite-relay-station in use on the air base transfers field data of unmanned aerial vehicles from Yemen to the U.S. or to other unmanned aerial vehicles and that the air base is a fundamental hub for the data transfer necessary to operate unmanned aerial vehicles in Yemen." As for the suit's demand that Germany prevent the relay station at Ramstein from facilitating drone strikes, the German government stated that it could not be expected to act "as a 'global public prosecutor' towards other sovereign states and punish alleged infringements outside of their own sovereign territory."

However, some legal scholars in Germany aren't satisfied with that response. They argue that if U.S. personnel based at Ramstein are involved in what the government considers an extrajudicial killing in a nondeclared war zone, they would not be entitled to immunity – at least not on German soil. The NATO Status of Forces Agreement explicitly grants German authorities the right to investigate members of the U.S. military suspected of having committed a crime. German prosecutors have shown little interest in pursuing such action. The German government position boils down to this: *We have asked the Americans if they are violating any agreements or laws and they have said no. Case closed.*

"What happens between the U.S., Ramstein, and the drones is a division of labor in different locations," said Wolfgang Kaleck, the head of the European Center for Constitutional and Human Rights, one of the organizations bringing the Yemen suit against the German government. "The German government doesn't ask tough questions because they obviously don't want to know what really happens."

Germany has figured prominently in the U.S. drone war from the very beginning. In 2000 the U.S. Air Force launched an initiative to explore arming drones, the same year that the CIA, contemplating the assassination of Osama bin Laden, began using unarmed Predators to try to track this high-value target. It was through this surveillance project that a scientist working with the CIA and the U.S. military devised a prototype of the system for remotely operating drones that endures to this day.

Originally called "split operations," the method involved drone pilots operating from Ramstein, while the actual aircraft would fly out of an airfield in Afghanistan's neighbor Uzbekistan. From there the drones could record live video over a complex near Kandahar, where bin Laden was suspected of residing. "They chose Ramstein because that was the most convenient place where they could be on a very secure location and still reach a satellite that had a footprint that covered Afghanistan," said Richard Whittle, author of the book *Predator: The Secret Origins of the Drone Revolution*. "And that worked." The successful development of the split operations was welcomed by those within the U.S. intelligence community who were pushing for authorities to assassinate bin Laden; it would make their mission easier to accomplish.

But plans to assassinate bin Laden with a Hellfire missile launched from a drone piloted from Ramstein hit a snag. "A Defense Department lawyer raised the issue that you couldn't pull the trigger from German soil under the U.S. Status of Forces Agreement without telling the German government you were going to do it and getting their permission," said Whittle. Fearing that the German government of Chancellor Gerhard Schroeder would reject the proposal or that the existence of the facility and the plot to kill bin Laden would leak, the CIA went back to the drawing board. "You have to remember, at that time the whole idea of assassinating Osama bin Laden had a different feel to it than it did later after 9/11," Whittle said. "He was barely known among the general public. The whole idea of the CIA running a targeted killing was entirely different, and there was a lot of hesitation."

The CIA considered moving the ground control station to a ship in the ocean or to another European location. But both of those scenarios would come with risks and technical complications. In the end the CIA decided to position pilots at a ground control station within CIA headquarters in Langley and use underwater fiber optic cables to facilitate lightning-fast communications between those pilots and the drones they would control. The cable to Germany was the artery connecting the pilots to the planes that would hunt bin Laden and other terror suspects. It would run from the United States to Ramstein, which would house a powerful satellite uplink that could hit satellites in Afghanistan. But the key was that the actual commands to deploy drones as weapons would be issued from American, not German soil, thus freeing the United States from the obligation to get the Germans' approval for the mission. The system was called "remote split operations."

Soon after taking office in 2009 President Obama authorized an expansion of the drone war, including opening new fronts in Somalia and Yemen. But the U.S. military discovered a gap in its satellite coverage. So in early 2009, after "an urgent call from the Pentagon's Joint staff," a commercial satellite provider, Intelsat, shifted its Galaxy-26 satellite from the United States to orbit over the Indian Ocean.[8] This repositioning allowed signals from U.S.-based drone operators to reach Galaxy-26 by using the relay station at Ramstein, facilitating the rapid expansion of the drone program.

Former drone sensor operator Brandon Bryant, who conducted operations in Yemen, Afghanistan, and Iraq, said that without Ramstein, the United States would either need to find another base in the area, with the ability to hit satellites in the Middle East and Africa, or place U.S. personnel much closer to the areas they were targeting. "Instead of being able to be [inside the United States] with their operations, they would have to do more line-of-sight stuff, more direct deployments, more people going over there rather than [operating] in the states," said Bryant, who has become an outspoken critic of the drone program. The United States, he charged, is "doing shady stuff behind the scenes, like using satellite and information technologies that, if able to continue being used, are going to just continue to perpetuate the drone war."

"Ramstein is the focal point for drone communications," said Dan Gettinger, codirector of the Center for the Study of the Drone at Bard College. "If the communications infrastructure didn't exist, the drone would be just a remote-control plane, a toy basically." It is "more important to the drone operations than the weapons a drone carries."

The top-secret slides show how embedded Ramstein has become in the drone war. They describe in detail the system by which a geolocating device affixed to the drone feeds back to a satellite and down to the station at Ramstein. The Gilgamesh platform, which *The Intercept* first reported on in February 2014, utilizes a device placed on the bottom of the drone. It operates as a fake cell phone tower, forcing individual mobile phones of targeted individuals to connect to it so that their location can be pinpointed and used in "find, fix and finish" missions. The slides show that Gilgamesh operations ran out of several sites, including in Djibouti, from which the United States has launched drone aircraft into Somalia and Yemen. The slides specify that drones are also equipped with the collection platform Airhandler, which relays data back to ground control stations via Ramstein.

Ramstein is not the only crucial U.S. military installation in Germany. A separate key facility, the European Technical Center (ETC), is an hour away, in Wiesbaden. According to a classified document provided by NSA whistleblower Edward Snowden, the ETC "is NSA's primary communications hub in that part of the world, providing communications connectivity, SIGINT collection, and data-flow services to NSAers, warfighters and foreign partners in Europe, Africa and the Middle East."[9]

In the top-secret drone architecture slide deck, the ETC is shown as having satellite links to Bagram Air Base in Afghanistan as well as a fiber optic connection to the NSA's counterterrorism facilities in Georgia, where many Gilgamesh operators supporting drone operations are based.

As the United States expands the global reach of its drones, Ramstein is poised to play a crucial role in new war frontiers. In June 2014 the U.S. Air Force awarded a contract to a major satellite provider, boasting that it "leverages our global satellite fleet to provide communications capability" for drones. The contract will support the operations of the Germany-based U.S. Africa Command. "Work will be performed at Ramstein Air Base, Germany and the western portion of Africa," the contract announcement states. In 2011 the U.S. Air Force requested $15 million to build a center similar to the Ramstein satellite facility at a U.S. military base in Sigonella, Italy. As of November 2014, according to a U.S. military contracting document, the project was still in a pre-solicitation stage. The request for funding of the station underlined the centrality of Ramstein to all current drone operations by asserting that the proposed site in Italy would "act as a back-up system to the Ramstein site to avoid single point of failure."

TOUCHDOWN

Hellfire missiles, the explosives fired from drones, are not always fired at people. In fact most drone strikes are aimed at phones. The SIM card provides a person's location; when turned on, a phone can become a deadly proxy for the individual being hunted.

When a night raid or drone strike successfully neutralizes a target's phone, operators call that a "touchdown."

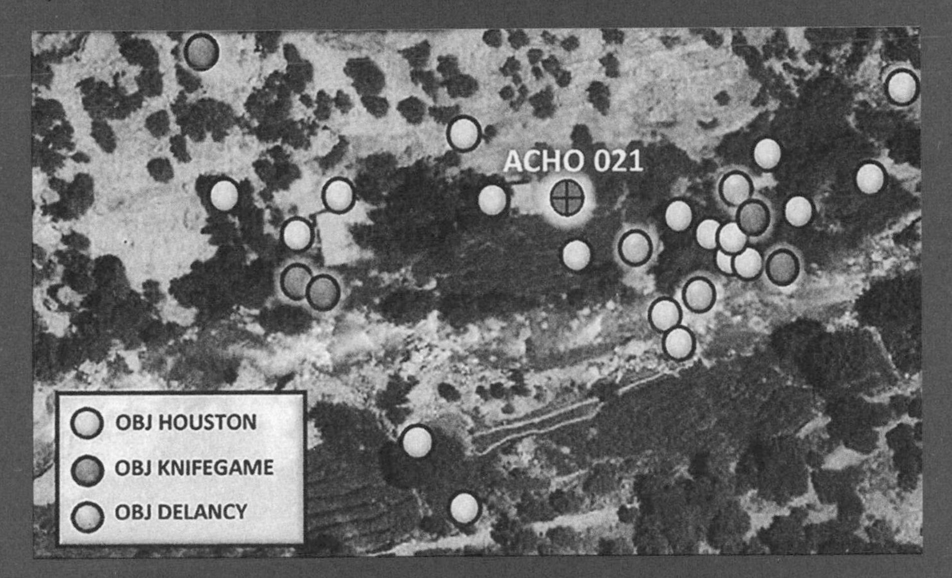

TARGET AFRICA

NICK TURSE

2005, May, 14
Lemonnier, Djibouti
Africa

2011, October, 28
Lemonnier, Djibouti
Africa

2012, June, 3
Lemonnier, Djibouti
Africa

2013, April, 8
Lemonnier, Djibouti
Africa

2007, February, 24
Lemonnier, Djibouti
Africa
2009, April, 16
Lemonnier, Djibouti
Africa
2012, August, 25
Lemonnier, Djibouti
Africa
2012, October, 2
Lemonnier, Djibouti
Africa
2013, October, 16
Lemonnier, Djibouti
Africa
2015, May, 23
Lemonnier, Djibouti
Africa

Morning dawns at length in Africa. The night has been long and dark. The opening day has a hopeful outlook and also an aspect of uncertainty. . . . For many years little colonies, trading-posts, and slave-marts have fringed its borders; but the vast interior has remained a blank.

Samuel Bartlett, *Historical Sketch of the Missions of the American Board in Africa*, 1880

Eradicating blank spaces on maps of the "dark continent" was an obsession of Western powers during the nineteenth-century scramble for Africa. Today a new scramble is under way to eradicate a different set of blank spots. Since 9/11 the U.S. military has engaged in a largely covert effort to extend its footprint across the continent with a network of mostly small and mostly low-profile camps. Some serve as staging areas for quick-reaction forces or bare-bones outposts where special ops teams can advise local proxies; some can accommodate large cargo planes, others only small surveillance aircraft.[1] All have one mission in common: to eradicate what the military calls the "tyranny of distance." These facilities allow U.S. forces to surveil and operate on larger and larger swaths of the continent – and, increasingly, to strike targets from drones and manned aircraft.

According to the 2013 Pentagon study on secret drone operations in Somalia and Yemen between January 2011 and summer 2012, a secretive unit known as Task Force 48-4 carried out a shadow war in the region. The task force, with its headquarters at Camp Lemonnier in Djibouti, operated from outposts in Nairobi, Kenya, and Sanaa, Yemen. The aircraft it used, both manned and remotely piloted, were based at airfields in Djibouti, Ethiopia, and Kenya, as well as ships off the coast of East Africa.

U.S. Africa Command (Africom), the umbrella organization for U.S. military activities on the continent, insists that it maintains only a "small footprint" in Africa and claims that Camp Lemonnier in Djibouti, a former French Foreign Legion outpost, is its only full-fledged base.[2] However, a number of new facilities have been opened in recent years, and even Defense Secretary Ashton Carter has acknowledged that Lemonnier serves as "a hub with lots of spokes out there on the continent and in the region."

One of those spokes can be found just ten kilometers southwest of Camp Lemonnier. After numerous mishaps and crashes, drone operations were moved from the camp to the more remote Chabelley Airfield in September 2013.[3] Predator drones have also been based in the cities of Niamey in Niger and N'Djamena in Chad, and Reaper drones have been flown out of Seychelles International Airport.[4] The ISR study also notes that, as of June 2012, there were two contractor-operated drones, one Predator and one Reaper, flying out of Arba Minch, Ethiopia.[5] Off the coast of East Africa a detachment equipped to dispatch a ScanEagle, a low-cost, low-tech drone used by the navy, or an MQ-8 Fire Scout, a remotely piloted helicopter, added to the regional array of surveillance assets, as did those associated with Armada Sweep, a ship-based system for collecting electronic communications.[6] (The NSA declined to comment on Armada Sweep.) Additionally two manned fixed-wing aircraft were based in Manda Bay, Kenya. More recent reports indicate that the military's Joint Special Operations Command is now working out of two bases in Somalia, one in Kismayo and the other in Baledogle.[7]

While generally austere, many of these bases, including the airfields in Chabelley and Manda Bay, have expanded in recent years, with more on the way.[8] In 2014, for example, Capt. Rick Cook, who at the time was chief of Africom's engineer division, mentioned the potential for a "base-like facility" that would be "semi-permanent" and "capable of air operations" in Niger.[9] The National Defense Authorization Act for fiscal year 2016, introduced in April 2015, requests $50 million for construction of an "Airfield and Base Camp at Agadez, Niger . . . to support operations in western Africa."[10]

Since 9/11 a multitude of other facilities – including staging areas, cooperative security locations, and forward operating locations – have also popped up (or been beefed up) in Burkina Faso, Cameroon, Central African Republic, Gabon, Ghana, Kenya, Mali, Senegal, South Sudan, and Uganda.[11] A 2011 report by Lauren Ploch, an analyst in African affairs with the Congressional Research Service, mentions

U.S. MILITARY DRONE AND SURVEILLANCE NETWORK IN AFRICA 2012–2015

	LOCATION	AIRCRAFT / PERSONNEL	CONFIRMED
1	Djibouti (Lemonnier)	10× MQ-1 (Predators), 4× MQ-9 (Reapers), 6× U-28, 2× P-3MS, 8× F-15E	2012
2	Arba Minch, Ethiopia	1× MQ-1 (Predator), 1x MQ-9 (Reaper)	2012
3	Manda Bay, Kenya	2× MFW (Medium Fixed-Wing)	2012
4	Nairobi, Kenya	Unknown	2012
5	Indian Ocean (on ships)	ScanEagle/Fire Scout detachment, Armada Sweep system	2012

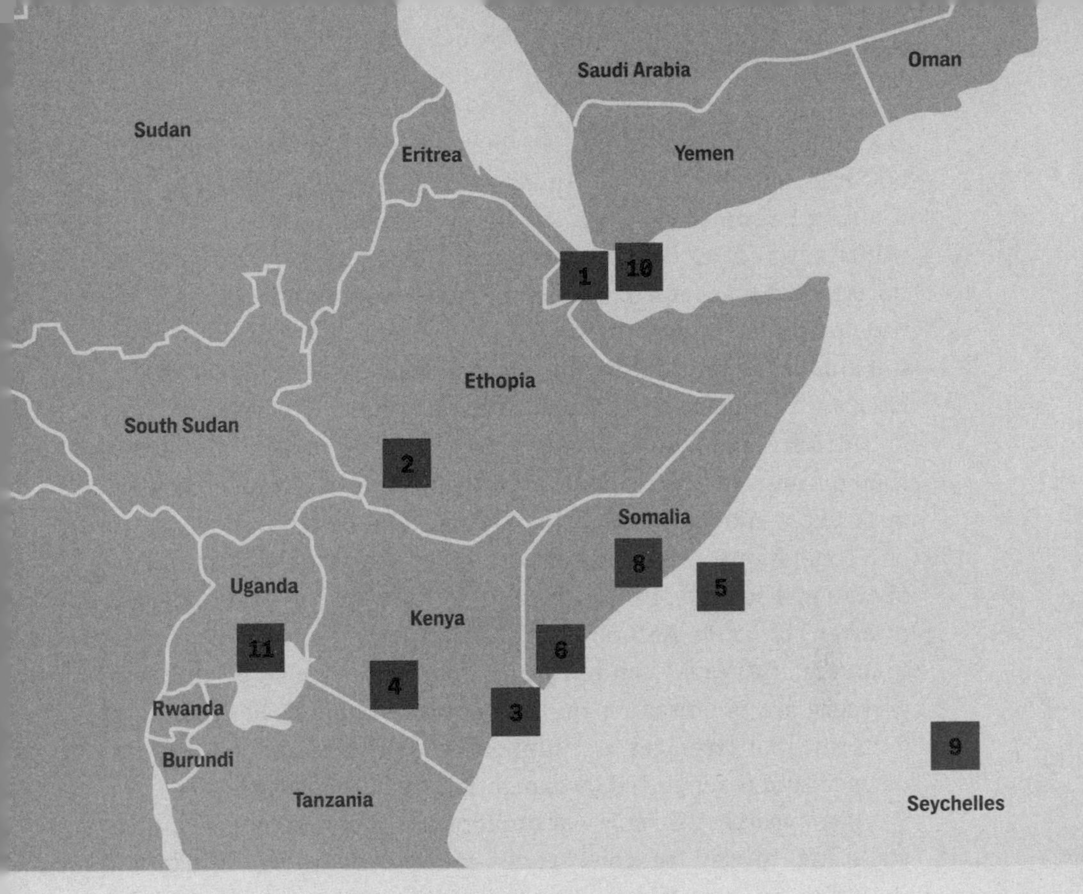

	LOCATION	AIRCRAFT / PERSONNEL	CONFIRMED
6	Kismayo, Somalia	Up to 40 people	2015
7	Niamey, Niger	MQ-1 (Predators)	2014
8	Baledogle, Somalia	30 to 40 people	2015
9	Seychelles	MQ-9 (Reapers)	2012
10	Djibouti (Chabelley)	MQ-9 (Reapers), MQ-1 (Predators)	2013
11	Entebbe, Uganda	PC-12 aircraft	2013
12	Ouagadougou, Burkina Faso	PC-12 aircraft	2013
13	N'Djamena, Chad	MQ-1 (Predators)	2014
14	Cameroon	MQ-1 (Predators), 90–300 people	2015

Sources: 1) ISR study; 2) ISR study; 3) ISR study; 4) ISR study; 5) ISR study; 6) Foreign Policy; 7) The Washington Post; 8) Foreign Policy; 9) The Washington Post; 10) The Washington Post; 11) The Washington Post; 12) The Washington Post; 13) The New York Times; 14) The Washington Post

U.S. military access to locations in Botswana, Namibia, São Tomé and Príncipe, Sierra Leone, Tunisia, and Zambia. According to Sam Cooks, a liaison officer with the Defense Logistics Agency, the U.S. military has struck twenty-nine agreements to use international airports in Africa as refueling centers. These locations are only some of the nodes in a growing network of outposts facilitating an increasing number of missions by the five thousand to eight thousand U.S. troops and civilians who annually operate on the continent.[12]

Africom and the Pentagon jealously guard information about their outposts in Africa, making it impossible to ascertain even basic facts like a simple count, let alone just how many are integral to JSOC operations, drone strikes, and other secret activities. "Due to operational security, I won't be able to give you the exact size and number," Lt. Cmdr. Anthony Falvo, an Africom spokesperson, wrote in an email. "What I can tell you is that our strategic posture and presence are premised on the concept of a tailored, flexible, light footprint that leverages and supports the posture and presence of partners and is supported by expeditionary infrastructure."

If you search Africom's website for news about Camp Lemonnier, you'll find myriad feel-good stories about green energy initiatives, the drilling of water wells, and a visit by country music star Toby Keith.[13] But that's far from the whole story. The base is a lynchpin for U.S. military action in Africa. "Camp Lemonnier is . . . an essential regional power projection base that enables the operations of multiple combatant commands," wrote Gen. Carter Ham in 2012, then the commander of Africom, in a statement to the House Armed Services Committee. "The requirements for Camp Lemonnier as a key location for national security and power projection are enduring."[14]

A map in the ISR study indicates that there were ten MQ-1 Predator drones and four larger, more far-ranging MQ-9 Reapers based at Camp Lemonnier in June 2012. There were also six U-28As, single-engine aircraft that conduct surveillance for special operations forces, and two P-3 Orions, four-engine turboprop aircraft originally developed for maritime patrols but since repurposed for use over African countries.[15] The map also shows the presence of eight F-15E Strike Eagles, manned fighter jets that are much faster and more heavily armed than drones. By August 2012 an average of sixteen drones and four fighter jets were taking off or landing there each day.[16]

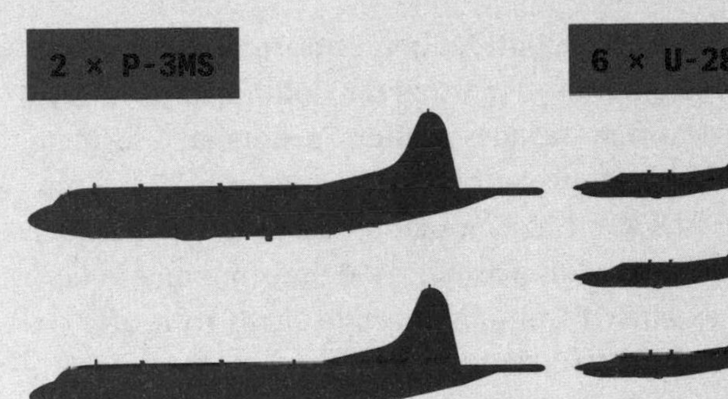

6 × U-28

8 × F-15E

U.S. AIRCRAFT BASED IN DJIBOUTI FOR TARGETED KILLING PROGRAM

4 × MQ-9 REAPER

10 × MQ-1 PREDATOR

Camp Lemonnier, a base in Djibouti, has been a focal point of the U.S. Military's drone operations in Africa and the Arabian Peninsula. A secret Pentagon study obtained by *The Intercept* lists the drones as well as surveillance and attack aircraft that operated out of Lemonnier in 2012 as part of the military's targeted killing program.

Located on the edge of Djibouti-Ambouli International Airport, Camp Lemonnier is also the headquarters for Combined Joint Task Force–Horn of Africa, which includes soldiers, sailors, and airmen, some of them members of special operations forces.[17] The camp, which also supports U.S. Central Command, has seen the number of personnel stationed there jump around 450 percent since 2002.[18] The base itself has expanded from eighty-eight acres to nearly six hundred and has been allocated or awarded more than $600 million for projects such as aircraft parking aprons, taxiways, and a major special operations compound.[19] In addition $1.2 billion in construction and improvements has been planned for the future.

As it grew, Camp Lemonnier became one of the most critical bases not only for the U.S. drone assassination campaign in Somalia and Yemen but also for U.S. military operations across the region.[20] The camp is so crucial to long-term military plans that in 2014 the United States signed a deal securing its lease until 2044, agreeing to hand over $70 million per year in rent – about double what it previously paid to the government of Djibouti.[21]

BASEBALL CARD

"Baseball cards" (BBCs) are the military's method for visualizing information; they are used to display data, map relationships between people, and identify an individual's so-called pattern of life.

This isn't quite what a baseball card looks like, but they are said to include much of the following information.

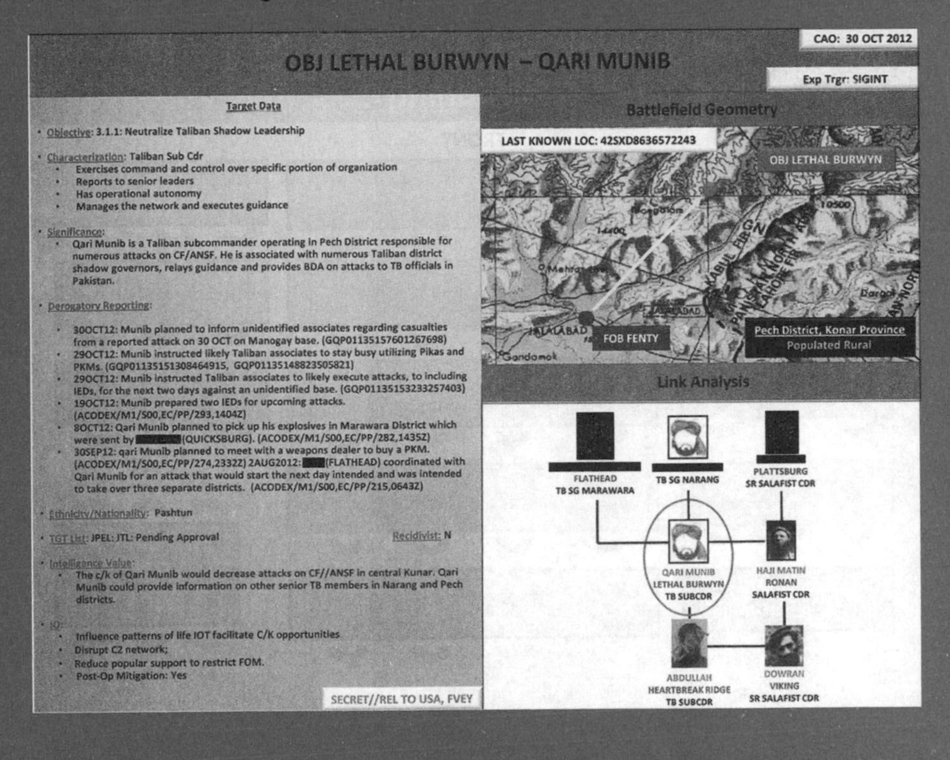

DEATH BY
METADATA
JEREMY SCAHILL
AND
GLENN GREENWALD

According to a former drone operator for the military's Joint Special Operations Command, the National Security Agency often identifies targets for drone strikes based on controversial metadata analysis and cell phone tracking technologies—an unreliable tactic that results in the deaths of innocent or unidentified people. Rather than confirming a target's identity with operatives or informants on the ground, the CIA or the U.S. military orders a strike based on the activity and location of the mobile phone a person is believed to be using.

The drone operator, who agreed to discuss the top-secret programs on the condition of anonymity, was a member of JSOC's High Value Targeting Task Force, which is charged with identifying, capturing, or killing terrorist suspects in Yemen, Somalia, Afghanistan, and elsewhere. His account is bolstered by top-secret NSA documents previously provided by NSA whistleblower Edward Snowden. It is also supported by former drone sensor operator Brandon Bryant, who has become an outspoken critic of the lethal operations in which he was directly involved in Iraq, Afghanistan, and Yemen.

In one tactic the NSA geolocates the SIM card or handset of a suspected terrorist's mobile phone, enabling the CIA and U.S. military to conduct night raids and drone strikes to kill or capture the individual in possession of the device. The former JSOC drone operator was adamant that the technology has been responsible for taking out terrorists and networks of people facilitating improvised explosive device attacks against U.S. forces in Afghanistan. But he also stated that innocent people have "absolutely" been killed as a result of the NSA's increasing reliance on the surveillance tactic. One problem, he explained, is that targets are increasingly aware of the NSA's reliance on geolocating and have moved to thwart the tactic. Some have as many as sixteen different SIM cards associated with their identity within the high-value target system. Others, unaware that their mobile phone is being targeted, lend their phone, with the SIM card in it, to friends, children, spouses, and family members. Some top Taliban leaders, knowing of the NSA's targeting method, have purposely and randomly distributed SIM cards among their units in order to elude their trackers. "They would do things like go to meetings, take all their SIM cards out, put them in a bag, mix them up, and everybody gets a different SIM card when they leave," the former drone operator said. "That's how they confuse us."

As a result, even when the agency correctly identifies and targets a SIM card belonging to a terror suspect, the phone may actually be carried by someone else, who is then killed in a strike. According to the former drone operator, the geolocation cells at the NSA that run the tracking program, known as Geo Cell, sometimes facilitate strikes without knowing whether the individual in possession of a tracked cell phone or SIM card is in fact the intended target of the strike. "Once the bomb lands or a night raid happens, you know that phone is there," he said. "But we don't know who's behind it, who's holding it. It's of course assumed that the phone belongs to a human being who is nefarious and considered an 'unlawful enemy combatant.' This is where it gets very shady."

The former drone operator said that he personally participated in drone strikes in which the identity of the target was known, but other, unknown people nearby were also killed. "They might have been terrorists," he said, "or they could have been family members who have nothing to do with the target's activities." What's more, he added, the NSA often locates drone targets by analyzing the activity of a SIM card rather than the actual content of the calls. Based on his experience, he has come to believe that the drone program amounts to little more than death by unreliable metadata. "People get hung up that there's a targeted list of people," he said. "It's really like we're targeting a cell phone. We're not going after people – we're going after their phones, in the hopes that the person on the other end of that missile is the bad guy."

The drone operator's account starkly contradicts the public posture of the Obama administration, which has repeatedly insisted that its operations kill terrorists with the utmost precision. In his speech at the National Defense University in 2014, President Obama declared that "before any strike is taken, there must be near-certainty that no civilians will be killed or injured – the highest standard we can set." He added, "By narrowly targeting our action against those who want to kill us and not the people they hide among, we are choosing the course of action least likely to result in the loss of innocent life."

But the increased reliance on phone tracking and other fallible surveillance tactics suggests that the opposite is true. In January 2014 the Bureau of Investigative Journalism, which uses a conservative methodology to track drone strikes, estimated that at least 273 civilians in Pakistan, Yemen, and Somalia were killed by unmanned aerial assaults during the first five years of the Obama administration.[1] A recent study conducted by a U.S. military adviser found that, during a single year in Afghanistan, where the majority of drone strikes have taken place, unmanned vehicles were ten times more likely than conventional aircraft to cause civilian casualties.

The NSA declined to answer questions about its role in drone strikes. Caitlin Hayden, a spokesperson for the National Security Council, also refused to discuss "the type of operational detail that, in our view, should not be published." In describing the administration's policy on targeted killings, Hayden would not say whether strikes are ever ordered without the use of human intelligence. She emphasized, "Our assessments are not based on a single piece of information. We gather and scrutinize information from a variety of sources and methods before we draw conclusions." Hayden felt free, however, to note the role that human intelligence plays after a deadly strike occurs: "After any use of targeted lethal force, when there are indications that civilian deaths may have occurred, intelligence analysts draw on a large body of information – including human intelligence, signals intelligence, media reports, and surveillance footage – to help us make informed determinations about whether civilians were in fact killed or injured."

The government does not appear to apply the same standard of care in selecting targets for assassination. The former JSOC drone operator estimates that the overwhelming majority of high-value target operations he worked on in Afghanistan relied on signals intelligence (SIGINT), based on the NSA's phone-tracking technology. "Everything they turned into a kinetic strike or a night raid was almost 90 percent that," he said. "You could tell, because you'd go back to the mission report and it will say 'This mission was triggered by SIGINT,' which means it was triggered by a geolocation cell."

In a July 2013 report the *Washington Post* relied exclusively on former senior U.S. intelligence officials and anonymous sources to herald the NSA's claims of effectiveness at geolocating terror suspects. Within the NSA, the paper reported, "a motto quickly caught on at Geo Cell: 'We Track 'Em, You Whack 'Em.'"[2] The *Post* article included virtually no skepticism about the NSA's claims and no discussion at all about how the unreliability of the agency's targeting methods results in the killing of innocents.

In fact, as the former JSOC drone operator recounts, tracking people by metadata and then killing them by targeting their SIM card is inherently flawed. The NSA "will develop a pattern," he said, "where they understand that this is what this person's voice sounds like, this is who his friends are, this is who his commander is, this is who his subordinates are. And they put them into a matrix. But it's not always correct. There's a lot of human error in that."

The JSOC operator's account is supported by another insider who was directly involved in the drone program. Brandon Bryant spent six years as a "stick monkey" – a drone sensor operator who controls the "eyes" of the U.S. military's unmanned aerial vehicles. By the time he left the air force in 2011, his squadron, which included a small crew of veteran drone operators, had been credited with killing 1,626 "enemies" in action. Bryant said that he has come forward because he is tormented by the loss of civilian life he believes he and his squadron may have caused. Today he is committed to informing the public about lethal flaws in the U.S. drone program.

Bryant describes the program as highly compartmentalized; drone operators taking shots at targets on the ground have little idea where the intelligence is coming from. "I don't know who we worked with," he said. "We were never privy to that sort of information. If the NSA did work with us, I have no clue." During the course of his career, Bryant said, many targets of U.S. drone strikes evolved their tactics, particularly in their handling of cell phones. "They've gotten really smart now and they don't make the same mistakes as they used to. They'd get rid of the SIM card and they'd get a new phone, or they'd put the SIM card in the new phone."

According to the former JSOC drone operator, and as classified documents obtained from Snowden confirm, the NSA doesn't just locate the cell phones of terror suspects by intercepting communications from cell phone towers and Internet service providers. The agency also equips drones and other aircraft with devices known as "virtual base-tower transceivers," creating, in effect, a fake cell-phone tower that can force a targeted person's device to lock onto the NSA's receiver without his or her knowledge. That, in turn, allows the military to track the cell phone to within thirty feet of its actual location, feeding the real-time data to teams of drone operators who conduct missile strikes or facilitate night raids.

The NSA geolocation system used by JSOC is known by the code name Gilgamesh. Under the program, a specially constructed device is attached to the drone. As the drone circles, the device locates the SIM card or handset that the military believes is used by the target. Relying on this method, said the former JSOC drone operator, means that the "wrong people" could be killed due to metadata errors, particularly in Yemen, Pakistan, and Somalia. "We don't have people on the ground [there] – we don't have the same forces, informants, or information coming in from those areas – as we do where we have a strong foothold, like we do in Afghanistan. I would say that it's even more likely that mistakes are made in places such as Yemen or Somalia, and especially Pakistan." As of May 2013, according to the former drone operator, President Obama had cleared sixteen people in Yemen and five in Somalia for targeting in strikes. Before a strike is green-lit, he said, there must be at least two sources of intelligence. The problem is that both of those sources often involve NSA-supplied data rather than human intelligence (HUMINT).

The former drone operator explained the "find, fix, finish" process of tracking and ultimately killing a targeted person: "Since there's almost zero HUMINT operations in Yemen, at least involving JSOC, every one of their strikes relies on signals and imagery for confirmation: signals being the cell phone lock, which is the 'find,' and imagery being the 'unblinking eye,' which is the 'fix.'" The "finish" is the strike itself. "JSOC acknowledges that it would be completely helpless without the NSA conducting mass surveillance on an industrial level. That is what creates those baseball cards you hear about," featuring potential targets for drone strikes or raids.

President Obama's authorizations for "hits" remain valid for sixty days. If a target cannot be located within that period, it must be reviewed and renewed. According to the former drone operator, it can take eighteen months or longer to move from intelligence gathering to getting approval to actually carrying out a strike in Yemen. "What that tells me is that commanders, once given the authorization needed to strike, are more likely to strike when they see an opportunity – even if there's a high chance of civilians being killed too – because in their mind they might never get the chance to strike that target again."

While drones are not the only method used to kill targets, they have become so prolific that they are now a standard part of U.S. military culture – so much so that they are often given nicknames. Among those used in Afghanistan, said the former JSOC drone operator, were "Lightning" and "Skyraider." The latter was also referred to as "Sky Raper" "because it killed a lot of people." When operators were assigned to "Sky Raper" it meant that "somebody was going to die. It was always sent to the most high-priority missions."

Similar to the Gilgamesh system used by JSOC, the CIA uses an NSA platform called Shenanigans. That operation – disclosed for the first time by *The Intercept* in February 2014 – utilizes a pod on aircraft that vacuums up massive amounts of data from any wireless routers, computers, smartphones, or other electronic devices that are within range. One top-secret NSA document provided by Snowden was written by a Shenanigans operator who described his March 2012 deployment to Oman, where the CIA has established a drone base. From almost four miles in the air he searched for communications devices believed to be used by al Qaeda in the Arabian Peninsula in neighboring Yemen. The mission was code-named Victorydance. "The VICTORYDANCE mission was a great experience," the operator wrote. "It was truly a joint interagency effort between CIA and NSA. Flights and targets were coordinated with both CIAers and NSAers. The mission lasted 6 months, during which 43 flights were flown." The mission, he added, "mapped the Wi-Fi fingerprint of nearly every major town in Yemen."

The NSA has played an increasingly central role in drone killings over the past five years. In one top-secret NSA document from 2010, the head of the agency's Strategic Planning and Policy Division of the Counterterrorism Mission Management Center recounted the history of the NSA's involvement in Yemen. Shortly before Obama took office, the document reveals, the agency began to "shift analytic resources to focus on Yemen." In 2008 the NSA had only three analysts dedicated to al Qaeda in the Arabian Peninsula in Yemen. By the fall of 2009 it had forty-five analysts and was producing "high quality" signals intelligence for the CIA and JSOC. In December 2009, utilizing the NSA's metadata collection programs, the Obama administration dramatically escalated U.S. drone and cruise missile strikes in Yemen. The first strike in the country known to be authorized by Obama targeted an alleged al Qaeda camp in the southern village of al-Majala. That strike, which included the use of cluster bombs, resulted in the deaths of fourteen women and twenty-one children.[3] It is not clear whether the strike was based on metadata collection; the White House has never publicly explained the strike or the source of the faulty intelligence that led to the civilian fatalities.

Another top-secret NSA document confirms that the agency "played a key supporting role" in the drone strike in September 2011 that killed U.S. citizen Anwar al Awlaki, as well as another American, Samir Khan. According to the 2013 Congressional Budget Justification, "The CIA tracked [Awlaki] for three weeks before a joint operation with the U.S. military killed" the two Americans in Yemen, along with two other people. When Brandon Bryant left his air force squadron in April 2011, the unit was aiding JSOC in its hunt for Awlaki. The CIA took the lead in the hunt for Awlaki after JSOC tried and failed to kill him in the spring of 2011.

According to Bryant, the NSA's expanded role in Yemen has only added to what he sees as the risk of fatal errors already evident in CIA operations. Referring to the CIA he said, "They're very nondiscriminate with how they do things, as far as you can see their actions over in Pakistan and the devastation that they've had there. It feels like they tried to bring those same tactics they used over in Pakistan down to Yemen. It's a repeat of tactical thinking instead of intelligent thinking."

Those within the system understand that the government's targeting tactics are fundamentally flawed. According to the former JSOC

(U) The Death of Anwar Nasser Aulaqi

(TS//NF) Anwar Nasser Aulaqi, a dual U.S./Yemeni citizen, regional commander for AQAP, and well-known extremist lecturer who preached at two U.S. mosques attended by some of the September 2001 hijackers, was killed in Yemen on 30 September 2011. The CIA tracked Aulaqi for three weeks before a joint operation with the U.S. military killed Aulaqi. The special operation killed four operatives, including Samir Khan, another American who played a key role in inspiring attacks against the U.S. Aulaqi's death represents another integrated CIA and military success in the counterterrorism fight.

A secret NSA document describes the assassination of Anwar al Awlaki, a U.S. citizen.

drone operator, instructors who oversee Gilgamesh training emphasize, "'This isn't a science. This is an art.' It's kind of a way of saying that it's not perfect." Yet the tracking pods mounted on the bottom of drones have facilitated thousands of "capture or kill" operations in Afghanistan, Iraq, Yemen, Somalia, and Pakistan since September 11. One top-secret NSA document provided by Snowden notes that by 2009, "for the first time in the history of the U.S. Air Force, more pilots were trained to fly drones . . . than conventional fighter aircraft," leading to a "'tipping point' in U.S. military combat behavior in resorting to air strikes in areas of undeclared wars," such as Yemen and Pakistan. The document continues, "Did you ever think you would see the day when the U.S. would be conducting combat operations in a country equipped with nuclear weapons without a boot on the ground or a pilot in the air?" Even NSA operatives seem to recognize how profoundly the agency's tracking technology deviates from standard operating methods of war.

An NSA document from 2005 poses this question: "What resembles 'LITTLE BOY' (one of the atomic bombs dropped on Japan during World War II) and as LITTLE BOY did, represents the dawn of a new era (at least in SIGINT and precision geolocation)?" Its reply: "If you answered a pod mounted on an Unmanned Aerial Vehicle (UAV) that is currently flying in support of the Global War on Terrorism, you would be correct."

Another document boasts that geolocation technology has "cued and compressed numerous 'kill chains' (i.e. all of the steps taken to find, track, target, and engage the enemy), resulting in untold numbers of enemy killed and captured in Afghanistan as well as the saving of U.S. and Coalition lives." The former JSOC drone operator, however, remains greatly disturbed by the unreliability of

(S) New Tactical Collection System Joins the War on Terrorism

FROM:
Technical Advisor, Target Reconnaissance and Survey (S316)
Run Date: 01/18/2005

DISTANTFOCUS pod is new system for tactical SIGINT and precision geolocation... first deployed in December (S)

(U//FOUO) What resembles "LITTLE BOY" (one of the atomic bombs dropped on Japan during World War II) and as LITTLE BOY did, represents the dawn of a new era (at least in SIGINT and precision geolocation)?

(S) If you answered a pod mounted on an Unmanned Aerial Vehicle (UAV) that is currently flying missions in support of the Global War on Terrorism, you would be correct.

Excerpt from a 2005 NSA document.

such methods. Like other whistleblowers, including Snowden and Chelsea Manning, he said that his efforts to alert his superiors to the problems were brushed off: "The system continues to work because, like most things in the military, the people who use it trust it unconditionally." When he would raise objections about intelligence that was "rushed" or "inaccurate" or "outright wrong," "the most common response I would get was 'JSOC wouldn't spend millions and millions of dollars, and man hours, to go after someone if they weren't certain that they were the right person.' There is a saying at the NSA: 'SIGINT never lies.' It may be true that SIGINT never lies, but it's subject to human error."

The government's assassination program is actually constructed, he said, to avoid self-correction. "They make rushed decisions and are often wrong in their assessments. They jump to conclusions, and there is no going back to correct mistakes." Because there is an ever-increasing demand for more targets to be added to the kill list, the mentality is "Just keep feeding the beast."

The killing of Awlaki, followed two weeks later by the killing of his sixteen-year-old son, Abdulrahman al Awlaki, also an American citizen, motivated Bryant to speak out. In October 2013 he appeared before a panel of experts at the United Nations that included the

UN's special rapporteur on human rights and counterterrorism, Ben Emmerson, who was conducting an investigation into civilians killed by drone strikes. Dressed in hiking boots and brown cargo pants, Bryant called for "independent investigations" into the Obama administration's drone program. "At the end of our pledge of allegiance, we say 'with liberty and justice for all,'" he told the panel. "I believe that should be applied to not only American citizens, but everyone that we interact with as well, to put them on an equal level and to treat them with respect."

Unlike those who oversee the drone program, Bryant took personal responsibility for his actions in the killing of Awlaki. "I was a drone operator for six years, active duty for six years in the U.S. Air Force, and I was party to the violations of constitutional rights of an American citizen who should have been tried under a jury. And because I violated that constitutional right, I became an enemy of the American people."

Later, in an interview, Bryant told us, "I had to get out because we were told that the president wanted Awlaki dead. And I wanted him dead. I was told that he was a traitor to our country. . . . I didn't really understand that our Constitution covers people, American citizens, who have betrayed our country. They still deserve a trial." The killing of Awlaki and his son continued to haunt Bryant. The younger Awlaki, Abdulrahman, had run away from home to try to find his dad, whom he had not seen in three years. But his father was killed before Abdulrahman could locate him. Abdulrahman was then killed in a separate strike two weeks later as he ate dinner with his teenage cousin and some friends. The White House has never explained the strike. "I don't think there's any day that goes by when I don't think about those two, to be honest," Bryant said. "The kid doesn't seem like someone who would be a suicide bomber or want to die or something like that. He honestly seems like a kid who missed his dad and went [to Yemen] to go see his dad."

In May 2013 President Obama acknowledged that "the necessary secrecy" involved in lethal strikes "can end up shielding our government from the public scrutiny that a troop deployment invites. It can also lead a president and his team to view drone strikes as a cure-all for terrorism." And that, said the former JSOC operator, is precisely what has happened. Given how much the government now relies on drone strikes – and given how many of those strikes are now dependent on metadata rather than human intelligence – the operator warns that political officials may view the geolocation program as more dependable than it really is. "I don't know whether or not President Obama would be comfortable approving the drone strikes if he knew the potential for mistakes that are there," he said. "All he knows is what he's told."

Whether or not Obama is fully aware of the errors built into the program of targeted assassination, he and his top advisers have repeatedly made clear that he himself directly oversees the drone operation and takes full responsibility for it. Obama once reportedly told his aides, "[It] turns out I'm really good at killing people."[4]

93←

BLINK

→121

A blink happens when a drone has to move and there isn't another aircraft to continue watching a target. According to the classified documents, blinks are a major challenge facing the military, which always wants to have a "persistent stare."

HOA ISR Orbits	**Finding**: A key factor in Find/Fix failures is the frequent inability to maintain 24/7 persistent stare on active mission areas, especially when ISR is massed for Finishes
	Recommendation: Support Combatant Command (CCMD) requirements for additional ISR orbits to help prevent "blinking" on HVIs during demand surges

The conceptual metaphor of surveillance is sight. Perfect surveillance would be like having a lidless eye. Much of what is seen by a drone's camera, however, appears without context on the ground. Some drone operators describe watching targets as "looking through a soda straw."

FIRING BLIND

CORA CURRIER
AND PETER MAASS

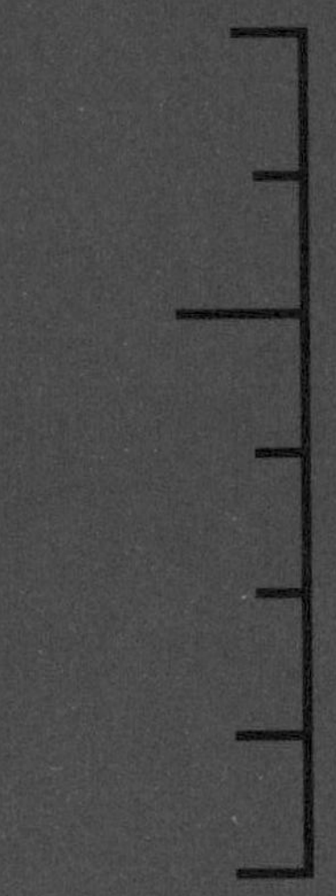

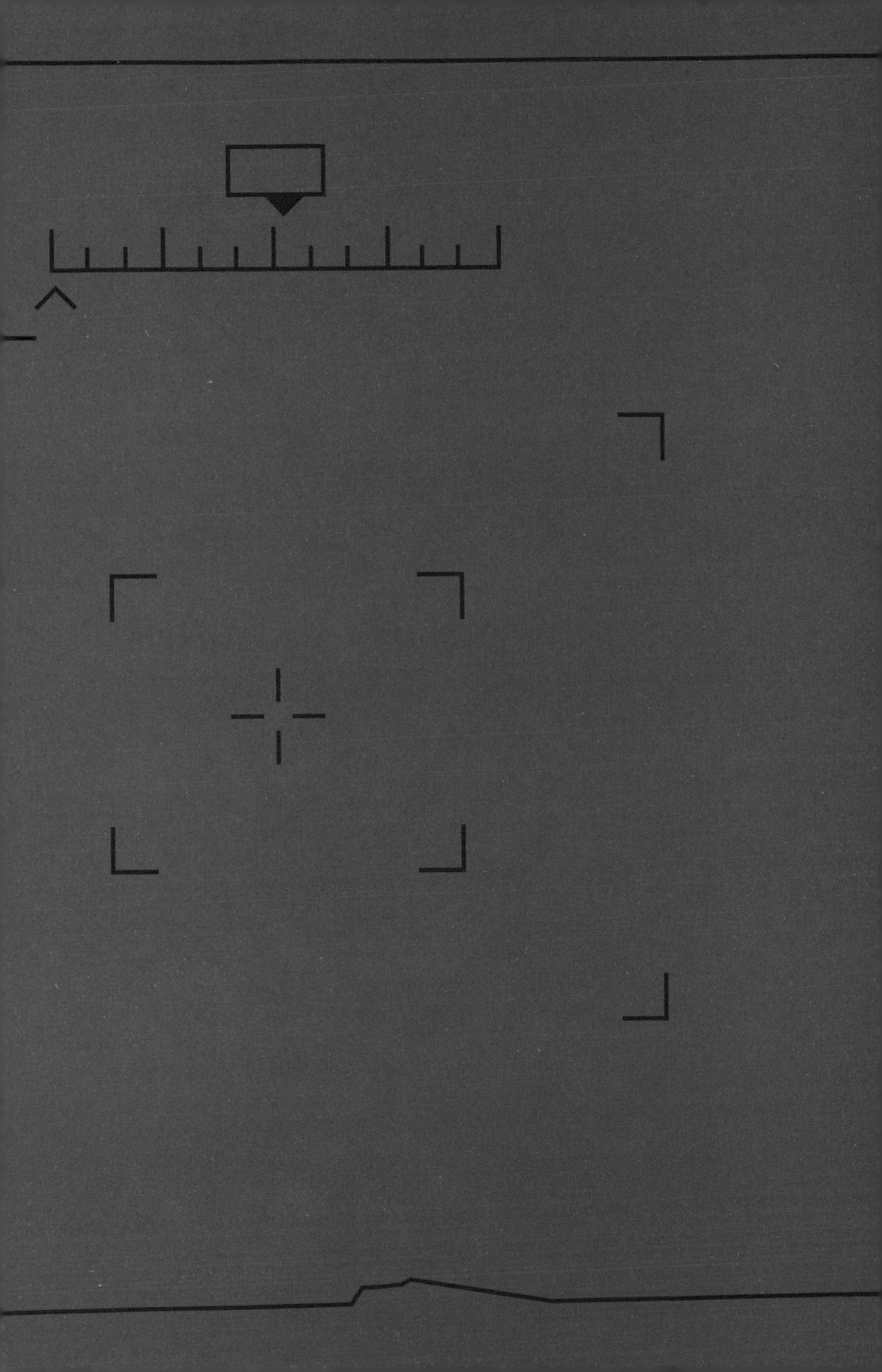

The irony of President Obama's drone war has been widely noted: an administration that wanted to stop torturing detainees and close down the controversial prison at Guantánamo Bay has wound up with an aerial killing campaign instead. There have been hundreds of drone strikes, killing thousands of people, during Obama's presidency, but details about the drone campaign, especially in areas outside Iraq and Afghanistan, have been difficult to obtain.

The Obama administration has portrayed drones as an effective and efficient weapon in the ongoing war with al Qaeda and other radical groups. Yet classified Pentagon documents, published by *The Intercept* in October 2015, reveal that the U.S. military has faced "critical shortfalls" in the technology and intelligence it uses to find and kill suspected terrorists in Yemen and Somalia. Those shortfalls stem from the remote geography of those countries and the limited American presence there. As a result the U.S. military has been overly reliant on signals intelligence from computers and cell phones, and the quality of those intercepts has been limited by constraints on surveillance flights in the region.

The documents are part of a study by a Pentagon Task Force on Intelligence, Surveillance, and Reconnaissance. They provide details about how targets were tracked for lethal missions carried out by the Joint Special Operations Command in Yemen and Somalia between January 2011 and summer 2012. When the study was circulated in

2013, the Obama administration was publicly floating the idea of moving the bulk of its drone program to the Pentagon from the CIA, and the military was eager to make the case for more bases, more drones, higher video quality, and better eavesdropping equipment.[1]

Yet by identifying the challenges and limitations facing the military's "find, fix, finish" operations in Somalia and Yemen – the cycle of gathering intelligence on, locating, and attacking a target – the conclusions of the ISR study would seem to undermine the Obama administration's claims of a precise and effective campaign and lend support to critics who have questioned the quality of intelligence used in drone strikes.

The study made specific recommendations for improving operations in the Horn of Africa, but a Pentagon spokesperson, Cmdr. Linda Rojas, declined to explain what, if any, measures had been taken in response to the study's findings. She said only, "As a matter of policy we don't comment on the details of classified reports."

THE TYRANNY OF DISTANCE

One of the most glaring problems identified in the ISR study was the U.S. military's inability to carry out full-time surveillance of its targets in the Horn of Africa and Yemen. Behind this problem lies "the tyranny of distance," a reference to the great distance that aircraft must fly to their targets from the main U.S. air base in Djibouti.

Surveillance flights are limited by fuel and, in the case of manned aircraft, the endurance of pilots. In contrast with Iraq, where more than 80 percent of "finishing operations" were conducted within 150

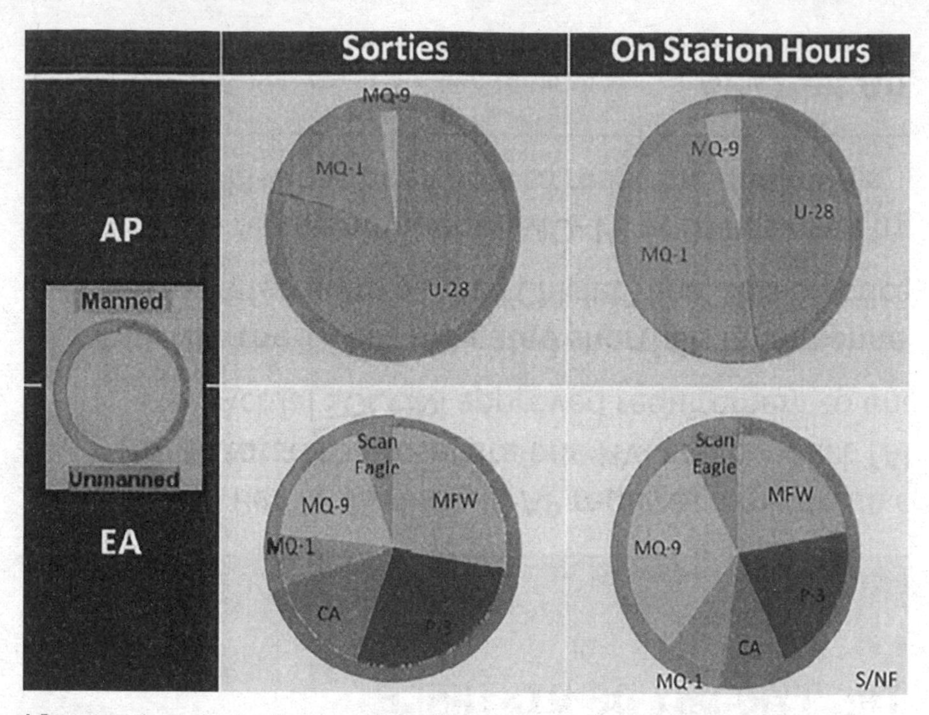

A Pentagon chart showing that as of June 2012 manned spy planes accounted for the majority of flights over Yemen, even though drones were more efficient, since they could spend more time over a target. Over Somalia the military used a mix of manned and unmanned aircraft.

kilometers of an air base, the study notes that "most objectives in Yemen are ~ 500 km away" from Djibouti, and those in "Somalia can be over 1,000 km [away]." The result is that drones and planes can spend half their air time in transit and not enough time conducting actual surveillance.

Compounding the tyranny of distance, the ISR study complained, was the fact that JSOC had too few drones in the region to meet the requirements mandated for carrying out a finishing operation. The military measures surveillance flights in orbits – meaning continuous, unbroken coverage of a target – and JSOC chronically failed to meet "minimum requirements" for orbits over Yemen; in the case of Somalia it had never met the minimum standards. On average fifteen flights a day by multiple aircraft relieving or complementing one another were needed to complete three orbits over Yemen.

The "sparse" available resources meant that aircraft had to "cover more potential leads – stretching coverage and leading to [surveillance] 'blinks.'" Because multiple aircraft needed to be "massed" over

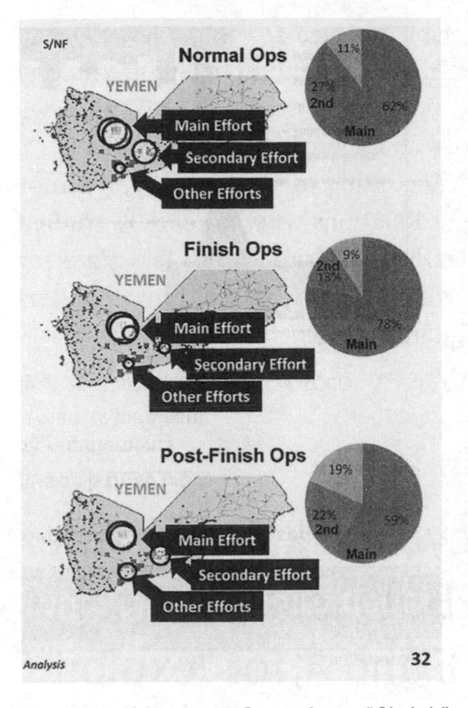

When the military was focused on a "finish" (kill) operation, drones were taken off the surveillance of other targets.

one target before a strike, surveillance of other targets temporarily ceased, thus breaking the military's ideal of a "persistent stare" or the "unblinking eye" of around-the-clock tracking.[2]

JSOC relied on manned spy planes to fill the orbit gap over Yemen. In June 2012 there were six U-28 spy planes in operation in East Africa and the Arabian Peninsula, as well as several other types of manned aircraft. The U-28s in Djibouti were "referred to as the 'Chiclet line,'" according to the ISR study, and "compounded Djiboutian air control issues" because of their frequent flights. Only in the summer of 2012, with the addition of contractor-operated drones based in Ethiopia and Fire Scout unmanned helicopters, did Somalia have the minimum number of drones commanders wanted. The number of Predator drones stationed in Djibouti doubled over the course of the study, and in 2013 the fleet was moved from the main U.S. air base, Camp Lemonnier, to the Chabelley airstrip because of overcrowding and a string of crashes.[3]

"Blinking" remained a concern, however, and the study recommended adding even more aircraft to the area of operations. Noting that political and developmental issues hampered the military's ability to build new bases, it suggested expanding the use of aircraft launched from ships. JSOC already made use of Fire Scout helicopter drones and small ScanEagle drones off the coast of Somalia, as well as "Armada Sweep," which a 2011 document from the National Security Agency, provided by former contractor Edward Snowden, describes as a "ship-based collection system" for electronic communications data. (The NSA declined to comment on Armada Sweep.)

Lt. Gen. Michael Flynn, who was head of the Defense Intelligence Agency from July 2012 to August 2014, said in an interview that the surveillance requirements he had outlined for tracking al Qaeda while in office had never been met. "We end up spending money on other stupid things instead of actually the capabilities that we need," he said. "This is not just about buying more drones. It's a whole system that's required." According to Micah Zenko, a senior fellow at the Council on Foreign Relations who has closely studied the drone war, resource constraints in Africa "mean less time for the persistent stare that counterterrorism analysts and commanders want and got used to in the Afghanistan-Pakistan theater."

FIND, FIX, FINISH

The "find, fix, finish" cycle is known in the military as FFF, or F3. But just as critical are two other letters, E and A, for "exploit and analyze," referring to the use of materials collected on the ground and in detainee interrogations.

F3EA became doctrine in counterinsurgency campaigns in Iraq and Afghanistan in the mid-2000s. Gen. Stanley McChrystal, the former commander of JSOC, wrote in his memoir that the simplicity of those "five words in a line . . . belied how profoundly it would drive our mission."[4] In 2008 Flynn, who worked closely with McChrystal before moving to the Defense Intelligence Agency, wrote, "Exploit-Analyze starts the cycle over again by providing leads, or start points, into the network that could be observed and tracked using airborne ISR."[5] Deadly strikes thus truncate the "find, fix, finish" cycle without exploitation and analysis – precisely the components that were lacking in the drone campaign waged in East Africa and Yemen. That shortfall points to one of the contradictions at the heart of the drone program in general: assassinations are intelligence dead ends.

The ISR study shows that after a "kill operation" there is typically nobody on the ground to collect written material or computers in

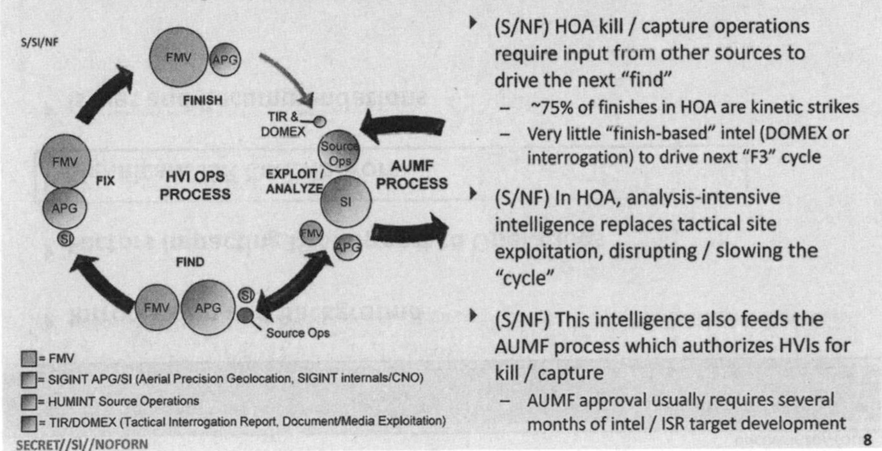

A slide from the ISR study notes that deadly strikes in Yemen and Somalia reduce the amount of intelligence for future operations.

the target's house or the phone on his body, or capture suspects and ask questions. Yet collection of on-the-ground intelligence of that sort – referred to as DOMEX, for "document and media exploitation," and TIR, for "tactical interrogation report" – is invaluable for identifying future targets.

Stating that 75 percent of operations in the region were strikes and that "kill operations significantly reduce the intelligence available from detainees and captured material," the study recommended an expansion of "capture finishes via host-nation partners for more 'finish-derived' intelligence." One of the problems with that scenario, however, is that security forces in host nations like Yemen and Somalia are profoundly unreliable and have been linked to a wide variety of abuses, including the torture of prisoners.

In a 2014 report by the Stimson Commission on U.S. Drone Policy, retired general John Abizaid and former Defense Department official Rosa Brooks wrote that the "enormous uncertainties" of drone warfare are "multiplied further when the United States relies on intelligence and other targeting information provided by a host nation

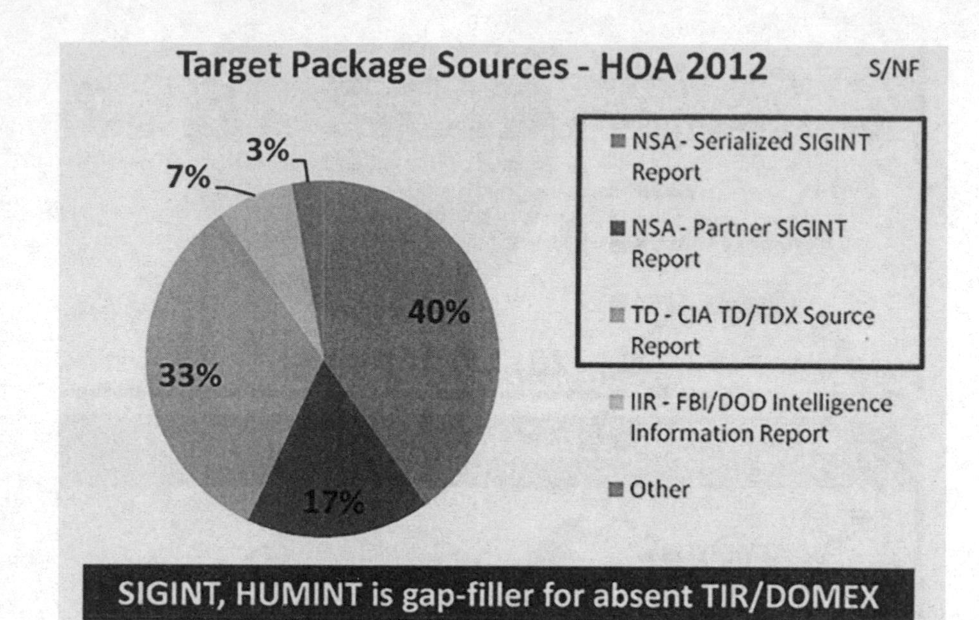

The military relies heavily on intelligence from electronic communications, much of it provided by foreign governments, but acknowledges that the information is "neither as timely nor as focused as tactical intelligence."

government: How can we be sure we are not being drawn into a civil war or being used to target the domestic political enemies of the host state leadership?"[6]

In 2011, for example, U.S. officials told the *Wall Street Journal* that they had inadvertently killed a local governor because Yemeni officials didn't tell them he would be present at a gathering of al Qaeda figures hit by a drone.[7] "We think we got played," one official said. (The Yemeni government disputed the report.) Despite such outcomes, the drone program has relied heavily on intelligence from other countries. One slide in the ISR study describes signals intelligence as coming "often from foreign partners," and another, titled "Alternatives to Exploit/Analyze," states, "In the reduced access environment, national intelligence partners often have the best information and access."

One way to increase the reliability of host-nation intelligence is to be directly involved in its collection, but this can be risky for soldiers on the ground. The ISR study called for "advance force operations," including "small teams of special force advisors," to work with foreign forces to capture combatants, interrogate them, and

A man walks past destroyed buildings in Zinjibar, capital of Abyan Province in southern Yemen, on December 5, 2012.

seize any written material or electronic devices they possess. According to public special operations guidelines, advance force operations "prepare for near-term" actions by planting tracking devices, conducting reconnaissance missions, and staging for attacks.[8] The ISR study did not specify an optimum number of advisers, where they should be based, or how exactly they should be involved in capture or interrogation operations.

Although the ISR study dates from 2013, Special Operations Commander Joseph Votel echoed its findings in July 2015. Votel reported that his troops were working closely with African Union forces and the Somali government to battle al Shabaab. He explained, "We get a lot more [intelligence] . . . when we actually capture somebody or we capture material than we do when we kill someone."[9]

THE POVERTY OF SIGNALS INTELLIGENCE

With limited ability to conduct raids or seize materials from targeted individuals in Yemen and Somalia, JSOC relied overwhelmingly on monitoring electronic communications to discover and ultimately locate targets. However, the ISR study states bluntly that

SIGINT is an inferior form of intelligence, although it accounted for more than half the intelligence collected on targets, much of it coming from foreign partners. The rest originated with human intelligence, primarily obtained by the CIA. "These sources," the study notes, "are neither as timely nor as focused as tactical intelligence" from interrogations or seized materials. Making matters worse, the documents refer to "poor" and "limited" capabilities for collecting SIGINT, implying a double bind in which kill operations were reliant on sparse amounts of inferior intelligence.

The disparity with other areas of operation was stark, as a chart contrasting cell data makes clear: in Afghanistan there were 8,900 cell data reports each month, compared to 50 for Yemen and 160 for Somalia. Despite that, another chart shows SIGINT comprised more than half the data sources that went into developing targets in Somalia and Yemen in 2012.

Lt. Gen. Flynn said in an interview that there was "way too much reliance on technical aspects [of intelligence], like signals intelligence, or even just looking at somebody with unmanned aerial vehicles. . . . I could get on the telephone from somewhere in Somalia, and I know I'm a high-value target, and say in some coded language, 'The wedding is about to occur in the next twenty-four hours.' That could put all of Europe and the United States on a high-level alert, and it may be just total bullshit. SIGINT is an easy system to fool, and that's why it has to be validated by other INTs, like HUMINT. You have to ensure that the person is actually there at that location because what you really intercepted was the phone."

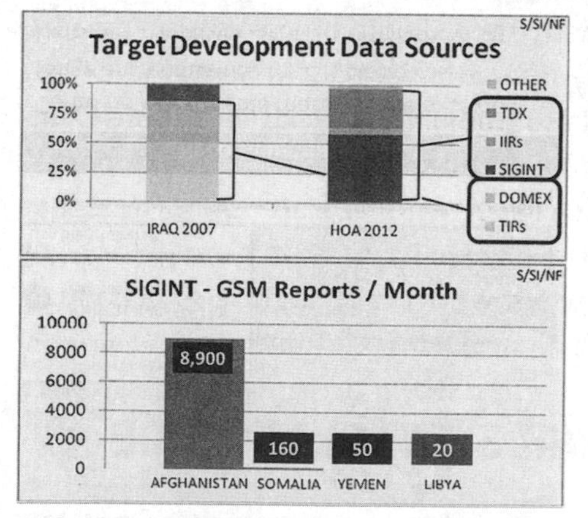

Cell phone data was critical for finding and identifying targets, yet a chart from the ISR study shows that the military had far less information in Yemen and Somalia than it was accustomed to having in Afghanistan.

In addition to using SIGINT to identify and find new targets, the documents detail how military analysts also relied on such intelligence to make sure they had the correct person in their sights and to estimate the harm to civilians before a strike. After locating a target, usually by his cell phone or other electronics, analysts would

SECRET//SI//NOFORN

ISR Platforms and Capabilities

Current ISR Systems used in HOA Small-Footprint Operations

System	Sensor					Platform				# of Aircraft in theater (as of 30 June 2012)	Armed for Operations (X)	Manned (M) Or Unmanned (U)
	FMV	HD-FMV	PTT COMINT	DNR COMINT	APG	Time On Station (hours) - Mogadishu	Time On Station (hours) - Sana'a	Cruise Speed (KIAS)	Max Endurance (Hours)			
P-3 MS	2		X	X	X	4	5	228	12	2		M
Medium Fixed Wing (MFW)	X		X	X	X	4	n/a	unk	8	2	X	M
U-28	2		X		X	--	3	270	5	6	X	M
MQ-1 Predator	1		X		X	6	12	70-90	20	6	X	U
MQ-9 Reaper	1	X	X		X	9	10	175	14	4	X	U
Scan Eagle	X					13	n/a	55	15	1 USN Det		U
MC-12 Liberty (Ext'd Range)	1		X		X	2(4)	4 (6)	300	6 (8)			M

SECRET//SI//NOFORN Red text denotes capabilities not in theater

- (S/NF) The PID-providing phenomenologies, HD-FMV and DNR COMINT, are largely absent from ISR systems operating in HOA
  - Not all MQ-9s have HD-FMV
  - MFW platforms currently only fly in Somalia
  - P-3 MS is a low-density / high-demand platform currently not in Theater

SECRET//SI//NOFORN *Source: HOA Orbit Tracker, as of June,2012 & aircraft spec sheets ; IBM Analysis* 39

A chart comparing the surveillance capabilities of the various drones and aircraft flying over Yemen and Somalia in 2012.

study video feeds from surveillance aircraft "to build near-certainty via identification of distinguishing physical characteristics."

A British intelligence document on targeted killing in Afghanistan, which was among the Snowden files, describes a similar process of "monitoring a fixed location, and tracking any persons moving away from that location, and identifying if a similar pattern is experienced through SIGINT collect." The document explains that "other visual indicators may be used to aid the establishment of [positive identification]," including "description of clothing" or "gait." After a shot, according to the British document and case studies in the Pentagon's ISR report, drones would hover to determine if their target had been hit, collecting video and evidence of whether the cell phone had been eliminated. (The British intelligence agency, GCHQ, declined to comment on the document.)

Yet according to the ISR study, the military faced "critical shortfalls of capabilities" in the technologies enabling that kind of precise surveillance and poststrike assessment. At the time of the study only some of the Reaper drones had high-definition video, and most of the aircraft over the region lacked the ability to collect "dial number recognition" data. The study cites these shortcomings as an explanation for the low rate of successful strikes against the targets on the military's kill list in Yemen and Somalia, especially in comparison with rates in Iraq and Afghanistan. It presents the failings primarily as an issue of efficiency, with little mention of the possible consequence of bad intelligence leading to killing the wrong people.

Drones are not magical. They have to take off from somewhere. Increasingly that somewhere is on the continent of Africa.

But where exactly?

As of 2012, the Joint Special Operations Command (JSOC) had bases in Djibouti, Kenya, and Ethiopia. JSOC operated eleven Predators and five Reaper drones over the Horn of Africa and Yemen.

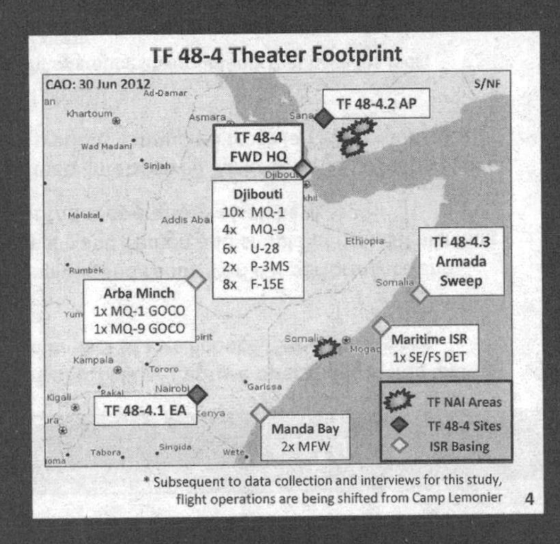

After crashing multiple Predator drones near Camp Lemonnier in Djibouti, the U.S. military moved operations to a more remote airstrip in Chabelley, Djibouti.

Here's a snapshot of how the United States views its surveillance capabilities on the continent of Africa more broadly.

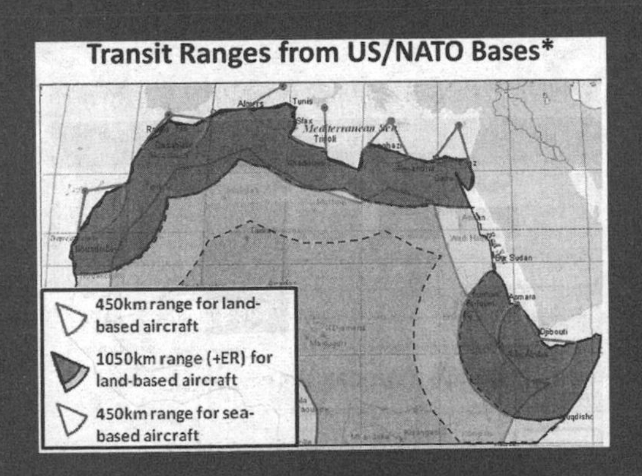

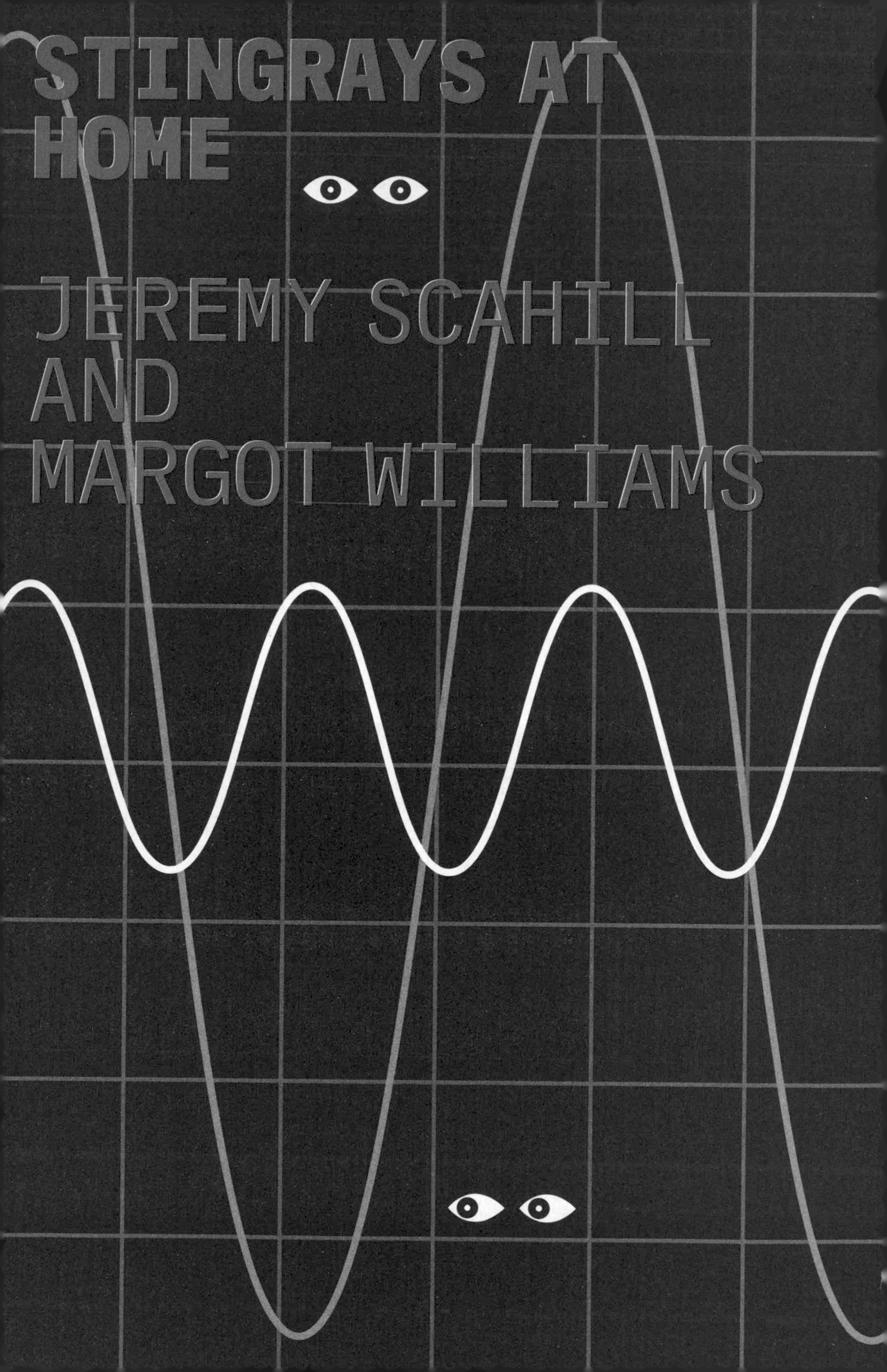
STINGRAYS AT HOME
JEREMY SCAHILL
AND
MARGOT WILLIAMS

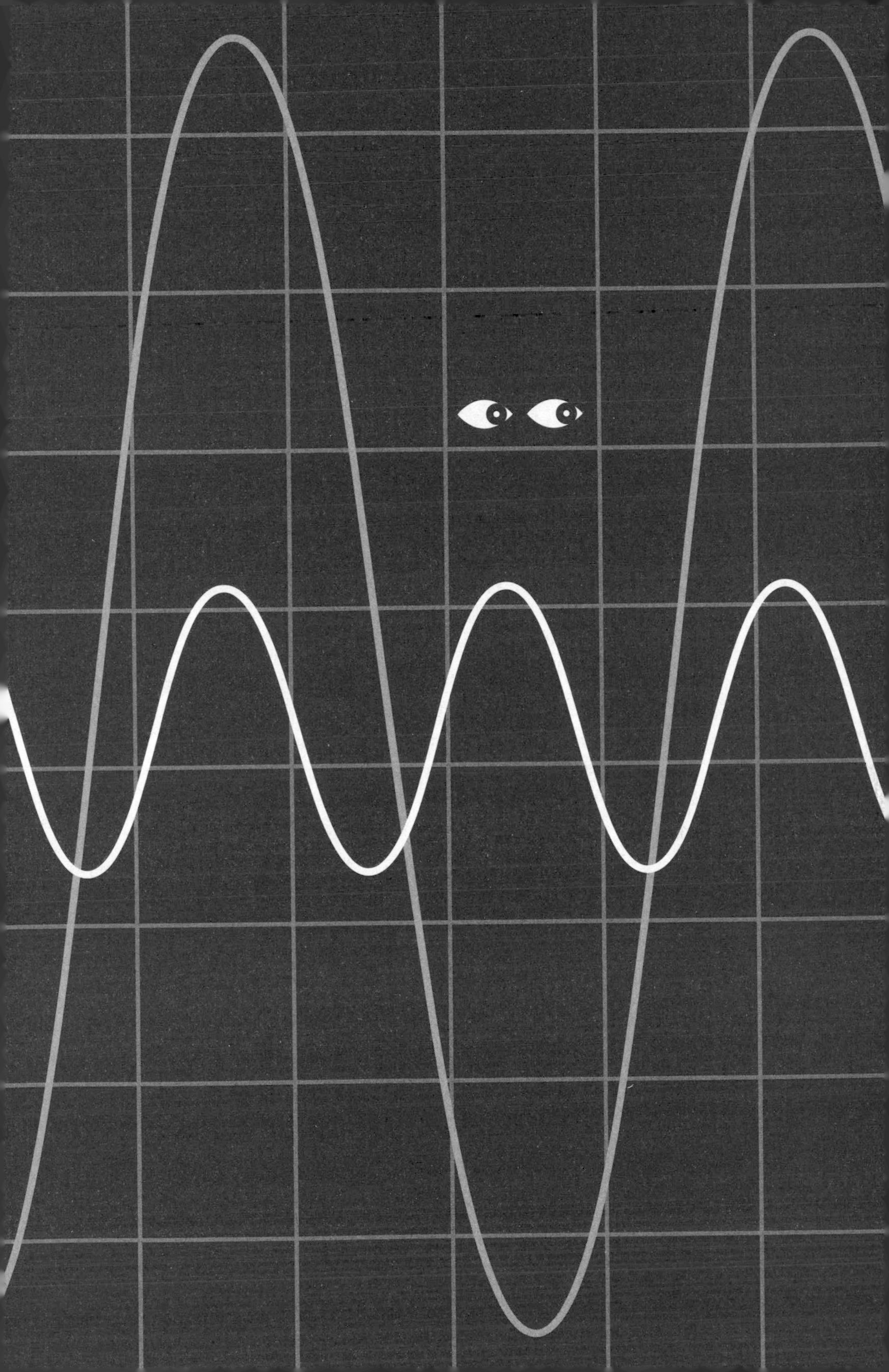

In December 2015 *The Intercept* published a secret internal U.S. government catalogue of dozens of cell phone surveillance devices used by the military and intelligence agencies.[1] The document, thick with previously undisclosed information, offers rare insight into the spying capabilities of federal law enforcement and local police inside the United States.

The catalogue, which was obtained from a source within the intelligence community concerned about the militarization of domestic law enforcement, includes details on the Stingray, a well-known brand of surveillance gear, as well as Boeing "dirt boxes" and dozens of more obscure devices that can be mounted on vehicles, drones, and piloted aircraft. Some are designed to be used at static locations, while others can be discreetly carried by an individual. They have names like Cyberhawk, Yellowstone, Blackfin, Maximus, Cyclone, and Spartacus. In the catalogue the NSA is listed as the vendor of one device; another was developed for use by the CIA; and another was developed for a special forces requirement. Nearly a third of the entries pertain to equipment that seem never to have been publicly described.

A few of the devices can house a "target list" of as many as ten thousand unique phone identifiers, and most can be used to geolocate people. Some have more advanced capabilities, like eavesdropping on calls and spying on SMS messages. Two systems, apparently designed for use on captured phones, are touted as having the ability to extract media files, address books, and notes, and one can retrieve deleted text messages.

Above all, the catalogue represents a trove of details on surveillance devices developed for military and intelligence purposes but increasingly used by law enforcement agencies to spy on people and convict them of crimes. The mass shooting in December 2015 in San Bernardino, California, which President Obama labeled "an act of terrorism," prompted calls for state and local police forces to beef up their counterterrorism capabilities, a process that has historically involved adapting military technologies to civilian use.[2] Meanwhile civil liberties advocates and others are increasingly alarmed at cell phone surveillance devices being used domestically and have called for a more open and informed debate on the trade-off between security and privacy – despite a virtual blackout by the federal government on any information about the specific capabilities of the gear.

"We've seen a trend in the years since 9/11 to bring sophisticated surveillance technologies that were originally designed for military use, like Stingrays or drones or biometrics, back home to the United States," said Jennifer Lynch, a senior staff attorney at the Electronic Frontier Foundation (EFF), which has waged a legal battle challenging the use of cell phone surveillance devices domestically. "But using these technologies for domestic law enforcement purposes raises a host of issues that are different from a military context."

Many of the devices in the catalogue, including the Stingrays and dirt boxes, are cell-site simulators, which operate by mimicking cell phone towers. When a phone connects to the spoofed network, it transmits a unique identification code and, through the characteristics of its radio signals when they reach the receiver, information about the phone's location. There are also indications that cell-site simulators may be able to monitor calls and text messages.[3]

The catalogue lists each device along with guidelines about how its use must be approved; the answer is usually via the "Ground Force Commander" or under one of two titles in the U.S. code governing military and intelligence operations, including covert action. But domestically, critics like Lynch say, the devices have been used in a way that violates the constitutional rights of citizens, including the Fourth Amendment prohibition on illegal search and seizure. The devices have regularly been used without warrants, or with warrants that critics call overly broad. Judges and civil liberties groups alike have complained that the devices are used without full disclosure of how they work, even within court proceedings. "Every time police drive the streets with a Stingray, these dragnet devices can identify and locate dozens or hundreds of innocent bystanders' phones," said Nathan Wessler, a staff attorney with the Speech, Privacy, and Technology Project of the American Civil Liberties Union.

The controversy over cell phone surveillance illustrates the friction that comes with deploying military combat gear in civilian life. The U.S. government has been using cell-site simulators for at least twenty years, but their use by local law enforcement is a more recent development.[4] The archetypical cell-site simulator, the Stingray, was trademarked by Harris Corp. in 2003 and initially used by the military, intelligence agencies, and federal law enforcement. Another company, Digital Receiver Technology, now owned by Boeing, developed dirt boxes – more powerful cell-site simulators – which gained favor among the NSA, CIA, and U.S. military as good tools for hunting down suspected terrorists. The devices can reportedly track more than two hundred phones over a wider range than the Stingray.

During the war on terror companies selling cell-site simulators to the federal government thrived. In addition to large corporations like Boeing and Harris, which clocked more than $2.6 billion in federal contracts in 2014, the catalogue obtained by *The Intercept* includes products from little-known outfits like Nevada-based Ventis, which appears to have been dissolved, and SR Technologies of Davie, Florida, which has a website that warns, "Due to the sensitive nature of this business, we require that all visitors be registered before accessing further information."[5] The catalogue is undated, but among the devices are models released in 2009, 2010, and 2011, as well as an indication that one was phased out in 2012. According to U.S. classification guidelines, the original secret designation date for the catalogue was 2006, but that is not the date of the creation of the catalogue or the specific devices.

As *The Intercept* reported in February 2014, the U.S. government eventually used cell-site simulators to target people for assassination in drone strikes. The secret catalogue contains slides outlining the capabilities of the Gilgamesh and Airhandler systems used by the military and CIA to "hunt high value targets" in Afghanistan and other war zones. But the CIA helped the U.S. Marshals Service use the technology at home too. For more than a decade, the *Wall Street Journal* revealed in March 2015, the agency worked with the marshals

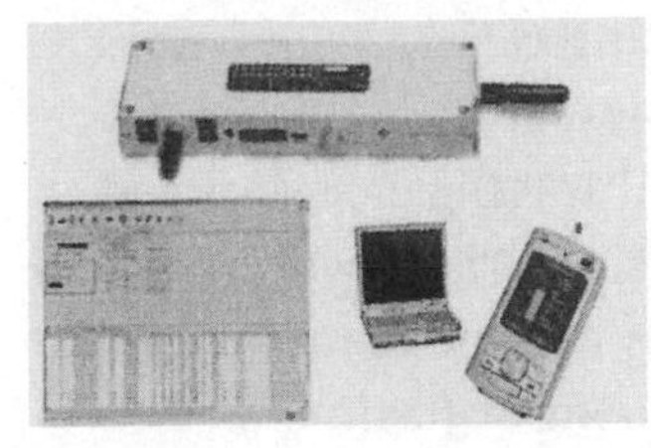

Blackfin I/II survey equipment.

"Can get you in a lot of trouble." Review by Nathan Wessler, staff attorney with the ACLU's Speech, Privacy, and Technology Project:

"From the maker of the Stingray, this device provides the added power to listen in on calls and read text messages. Also useful for kicking nearby phones off the network (you can choose between just blocking a single target phone or scrambling the signals of all phones in the area). Take note: Wiretapping calls and text messages requires a special 'super-warrant' signed by a judge. Playing around with a Blackfin without adequate court supervision can get you in a lot of trouble."

to deploy planes with dirt boxes attached to track mobile phones across the United States.[6]

After being used by federal agencies for years, cellular surveillance devices began to make their way into the arsenals of a small number of local police agencies. In 2007 Harris sought a license from the Federal Communications Commission to widely sell its devices to local law enforcement, and police flooded the FCC with letters of support.[7] "The text of every letter was the same. The only difference was the law enforcement logo at the top," said Chris Soghoian, the principal technologist at the ACLU, who obtained copies of the letters from the FCC through a Freedom of Information Act request.

The lobbying campaign was a success. As of early 2016, nearly sixty law enforcement agencies in twenty-three states were known to possess a Stingray or some form of cell-site simulator, though experts believe that number probably underrepresents the real total.[8] In some jurisdictions police use cell-site simulators regularly; the Baltimore Police Department, for example, has used Stingrays more than 4,300 times since 2007.[9] "Cell-site simulator technology has been instrumental in aiding law enforcement in a broad array of investigations, including kidnappings, fugitive investigations, and complicated narcotics cases," said Deputy Attorney General Sally Quillian Yates.

Police often cite the war on terror as their reason for acquiring such systems. Michigan State Police claimed their Stingrays would "allow the State to track the physical location of a suspected terrorist," although the ACLU later found that in 128 uses of the devices in 2014, none was related to terrorism.[10] In Tacoma, Washington, police claimed Stingrays could prevent attacks using improvised explosive devices, the roadside bombs that plagued soldiers in Iraq.[11] "I am not aware of any case in which a police agency has used a cell-site simulator to find a terrorist," said the EFF's Lynch. Instead "law enforcement agencies have been using cell-site simulators to solve even the most minor domestic crimes."

The Office of the Director of National Intelligence declined to comment on the catalogue or the devices described within it. The FBI, NSA, and U.S. military did not offer any comment after acknowledging our written requests. The Department of Justice "uses technology in a manner that is consistent with the requirements and protections of the Constitution, including the Fourth Amendment, and applicable statutory authorities," said Marc Raimondi, a Department of Justice spokesperson who previously spent six years working for Harris Corp., the manufacturer of the Stingray.

Cyberhawk Battlefield Data Recovery/SSE.

"More than enough data to map an entire social network."
Review by Jennifer Lynch, senior staff attorney, Electronic Frontier Foundation:

"Are you trying to break the next big criminal syndicate? If so, the Cyberhawk may be your go-to device. It's able to get data off over 79 cellphones, including saved and dialed numbers, SMS messages, pictures, calendar entries, and even sound files. That's more than enough data to map the entire social network of an organization—even if your targets use throwaway 'burner' phones. And the best thing about it? It uses software and components made right here in the United States. But better get a warrant to search those phones—in 2014, the Supreme Court said it's the law."

While interest from local police helped fuel the spread of cell-site simulators, funding from the federal government also played a role, incentivizing municipalities to buy more of the technology. In the years since 9/11 the United States has expanded its funding to provide military hardware to state and local law enforcement agencies via grants awarded by the Department of Homeland Security (DHS) and the Department of Justice (DOJ). There's been a similar pattern with Stingray-like devices.

"The same grant programs that paid for local law enforcement agencies across the country to buy armored personnel carriers and drones have paid for Stingrays," said the ACLU's Soghoian. "Like drones, license plate readers, and biometric scanners, the Stingrays are yet another surveillance technology created by defense contractors for the military, and after years of use in war zones, it eventually trickles down to local and state agencies, paid for with DOJ and DHS money."

In 2013 the Florida Department of Law Enforcement reported the purchase of two HEATR long-range surveillance devices and $3 million worth of Stingrays since 2008.[12]

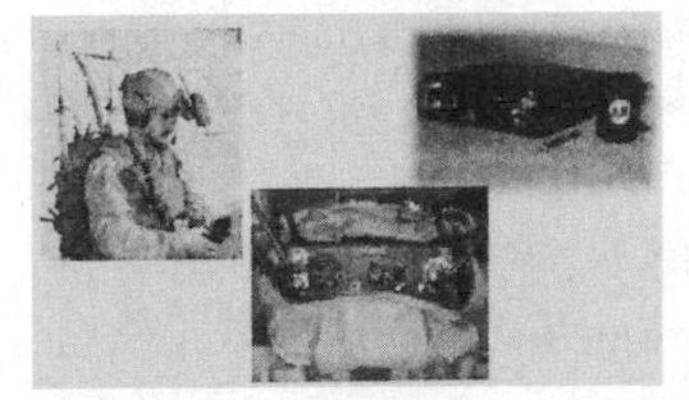

Spartacus II Ground-Based Geolocation (Vehicular).

"Can be easily moved from plane to car to body."
Review by Jennifer Lynch, senior staff attorney, Electronic Frontier Foundation:

"You'll feel like a powerful Greek gladiator with the Spartacus II. It's the smallest high-powered dual-band system on the market and can be moved easily from a plane to a car or even to your body—all without changing the system. While the $180,000 price tag might put it out of reach for smaller agencies, its cross-border capabilities could make it easy to acquire with DHS funding. And if it's used at the border, you might not even need to get a warrant before you use it."

In California, Alameda County and police departments in Oakland and Fremont are using $180,000 in Homeland Security grant money to buy Harris's Hailstorm cell-site simulator and the handheld Thoracic surveillance device, made by Maryland security and intelligence company Keyw.[13] As part of Project Archangel, which is described in government contract documents as a "border radio intercept program," the Drug Enforcement Administration has contracted with Digital Receiver Technology for over $1 million in surveillance box equipment. The Department of the Interior contracted with Keyw for more than half a million dollars of "reduced signature cellular precision geolocation."

Information on such purchases, like so much about cell-site simulators, has trickled out through freedom of information requests and public records. The capabilities of the devices are kept under lock and key – a secrecy that harks back to their military origins. When state or local police purchase the cell-site simulators, they are routinely required to sign nondisclosure agreements with the FBI that they may not reveal the "existence of and the capabilities provided by" the surveillance devices or share "any information" about the equipment with the public.[14]

Indeed while several of the devices in the military catalogue obtained by *The Intercept* are actively deployed by federal and local law enforcement agencies, according to public records judges have struggled to obtain details of how they work. Other products in the secret catalogue have never been publicly acknowledged; as a result any use by state, local, and federal agencies inside the United States is difficult to challenge.

"It can take decades for the public to learn what our police departments are doing, by which point constitutional violations may be widespread," the ACLU's Wessler said. "By showing what new surveillance capabilities are coming down the pike, these documents will help lawmakers, judges, and the public know what to look out for as police departments seek ever more powerful electronic surveillance tools."

Sometimes it's not even clear how much police are spending on Stingray-like devices because they are bought with proceeds from assets seized in drug busts and other operations under federal civil forfeiture law. Illinois, Michigan, and Maryland police forces have used asset forfeiture funds to pay for Stingray-type equipment. "The full extent of the secrecy surrounding cell-site simulators is completely unjustified and unlawful," said the EFF's Lynch. "No police officer or detective should be allowed to withhold information from a court or criminal defendant about how the officer conducted an investigation."

DRT 1101B Survey Equipment.

"Up to 10,000 targets."
Review by Jennifer Lynch, senior staff attorney, Electronic Frontier Foundation:

"Are you trying to monitor a huge political protest? Look no further than DRT. Nicknamed 'dirt boxes,' these devices can locate up to 10,000 targets and can process multiple analog and digital wireless devices all at the same time. They're even capable of intercepting and recording digital voice data. The best thing about the devices is the fact that no one may ever know you've used one. Just be careful—if your targets do figure out you've used a DRT box, and you haven't gotten a warrant, they may be able to convince a judge to throw out all the evidence you've collected on them after you used the device. You can mount DRT models like this one in an aircraft to fly over the crowd."

Judges have been among the foremost advocates for ending the secrecy around cell-site simulators, including by pushing back on warrant requests. At times police have attempted to hide their use of Stingrays in criminal cases, prompting at least one judge to throw out evidence obtained by the device. In 2012 a U.S. magistrate judge in Texas rejected an application by the Drug Enforcement Administration to use a cell-site simulator in an operation, saying that the agency had failed to explain "what the government would do with" the data collected from innocent people.

Law enforcement has responded with some limited forms of transparency. In September 2015 the Justice Department issued new guidelines for the use of Stingrays and similar devices, including that federal law enforcement agencies using them must obtain a warrant based on probable cause and must delete any data intercepted from individuals not under investigation.[15] Contained within the guidelines, however, is a clause stipulating vague "exceptional circumstances" under which agents could be exempt from the requirement to get a probable cause warrant. "Cell-site simulator technology has been instrumental in aiding law enforcement in a broad array of investigations, including kidnappings, fugitive investigations, and complicated narcotics cases," said Deputy Attorney General Sally Quillian Yates.

Meanwhile parallel guidelines issued by the Department of Homeland Security in October 2015 do not require warrants for operations on the U.S. border, nor do the warrant requirements apply to state and local officials who purchased their Stingrays with grants from the federal government, such as those in Wisconsin, Maryland, and Florida.[16] The ACLU, EFF, and several prominent members of Congress have said the federal government's exceptions are too broad and leave the door open to abuses. "Because cell-site simulators can collect so much information from innocent people, a simple warrant for their use is not enough," said Lynch, the EFF attorney. "Police officers should be required to limit their use of the device to a short and defined period of time. Officers also need to be clear in the probable cause affidavit supporting the warrant about the device's capabilities."

In November 2015 a federal judge in Illinois published a legal memorandum about the government's application to use cell tower spoofing technology in a drug-trafficking investigation. In his memo Judge Iain Johnston sharply criticized the secrecy surrounding Stingrays and other surveillance devices, suggesting that it made weighing the constitutional implications of their use extremely difficult. "A cell-site simulator is simply too powerful of a device to be used and the information captured by it too vast to allow its use without specific authorization from a fully informed court," he wrote. He added that Harris Corp. "is extremely protective about information regarding its device. In fact, Harris is so protective that it has been widely reported that prosecutors are negotiating plea deals far below what they could obtain so as to not disclose cell-site simulator information. . . . So where is one, including a federal judge, able to learn about cell-site simulators? A judge can ask a requesting Assistant United States Attorney or a federal agent, but they are tight-lipped about the device, too."[17]

The ACLU and EFF believe that the public has a right to review the types of devices being used in order to encourage an informed debate on the potentially far-reaching implications of the technology. The secret catalogue, said the ACLU's Wessler, "fills an important gap in our knowledge, but it is incumbent on law enforcement agencies to proactively disclose information about what surveillance equipment they use and what steps they take to protect Fourth Amendment privacy rights."

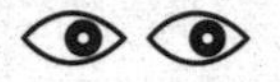

ORBITS

→153

The military worries about what it calls the "tyranny of distance" in its efforts to maximize orbits. An orbit, in this context, refers to a cycle of continuous, unbroken coverage of a target. Compared to the traditional battlefields of Iraq and Afghanistan, U.S. drones have to travel farther to reach their "named areas of interest," or NAIs, in Yemen and Somalia.

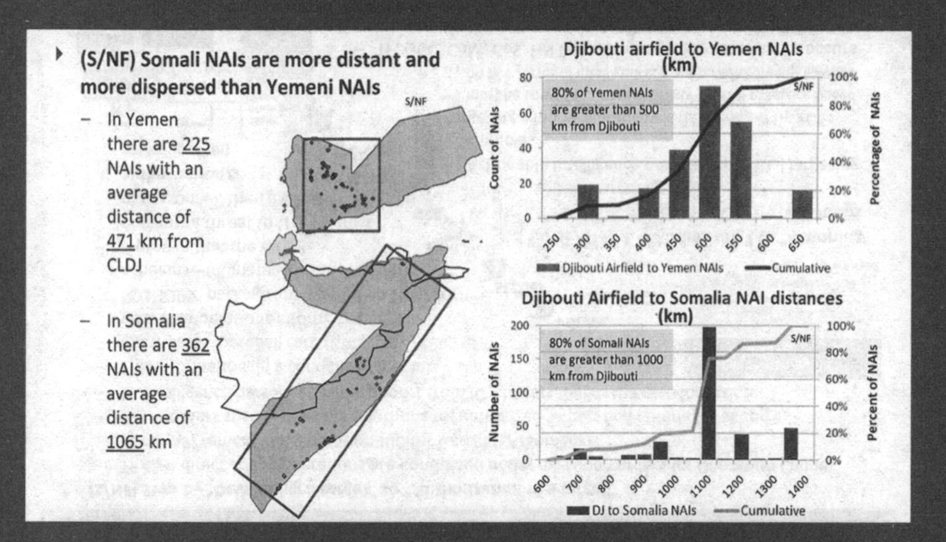

Here's where the United States appears to have "finished" people in Yemen.

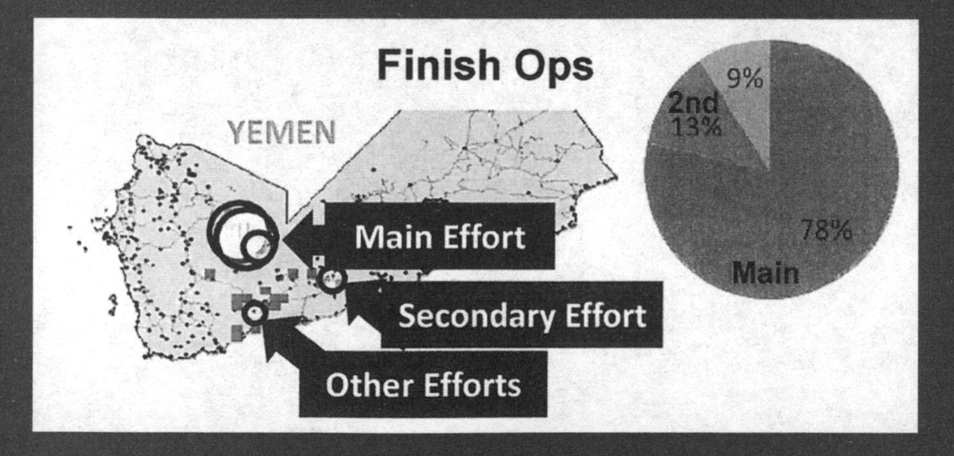

THE LIFE AND
DEATH OF
OBJECTIVE
PECKHAM
RYAN GALLAGHER

As he walked through the busy streets of London, Bilal el-Berjawi was glancing over his shoulder. Everywhere he went he suspected he was being followed. Within a few years, and four thousand miles away in remote Somalia, he would be dead, killed by a secret U.S. drone strike.

A small and stocky British Lebanese citizen with a head of thick dark hair, Berjawi had grown up much like any other young boy in the United Kingdom's capital city: attending school during the day and playing soccer with friends in his free time. But by his early twenties he was leading no ordinary life. He was suspected of having cultivated ties with senior al Qaeda militants in East Africa, his British citizenship was abruptly revoked, and he was placed on a U.S. kill list. In January 2012 Berjawi met his sudden end about ten miles northwest of Mogadishu, when a missile crashed into his white car and blasted it beyond recognition.

At the time of Berjawi's death the Associated Press reported that the missile strike targeting him had been carried out by a drone, citing an anonymous U.S. official.[1] *The Economist* criticized the secrecy surrounding the attack and questioned whether it had amounted to a "very British execution."[2] A classified U.S. document obtained and published by *The Intercept* in October 2015 shines new light on the circumstances surrounding Berjawi's death. It reveals that the U.S. government was monitoring him for at least five years as he traveled between London and Somalia; that he was targeted by a covert special operations unit running a fleet of more than two dozen drones, fighter jets, and other aircraft out of East Africa; and that cell phone surveillance facilitated the strike that killed him.

The document, a case study included in the 2013 report by the Pentagon's Intelligence, Surveillance, and Reconnaissance Task Force, does not mention Berjawi by name, instead referring to a target code-named "Objective Peckham." But it contains enough specific de-

A video produced by al Shabaab purports to show Berjawi's mangled vehicle in the aftermath of the drone strike that killed him on January 21, 2012.

tails about the target's movements and the time and place of the attack that killed him to confirm his identity beyond doubt. The story of Berjawi's life and death raises new questions about the British government's role in the targeted assassination of its own citizens and provides unique insight into covert U.S. military actions in the Horn of Africa and their impact on al Qaeda and its affiliate in the region, al Shabaab.

Berjawi – who was known by a variety of names, including Bilal Abul-Jariya, Abu Omar, and Abu Hafsa – spent his youth in the St. John's Wood district of northwest London, living in an apartment a short walk from Abbey Road Studios. He was a baby when his mother moved with him and his sister and brother to the United Kingdom. According to Berjawi's own account of his upbringing, which he provided in a lengthy interview with Cage, a London-based rights group, he was born in Lebanon in 1990 and moved to London the same year. But passport records uncovered by Ugandan media indicate that he may in fact have been born in September 1984, which would make him twenty-seven at the time of his death.

As a teenager Berjawi hung around with his friends on London's busy Edgware Road and frequented some of the shisha bars and Lebanese food stores scattered across the area. Tam Hussein, a former youth worker for a community organization in north London, met Berjawi for the first time around 2003, when Berjawi was sixteen or seventeen at the time, according to Hussein. "He was a good kid back then," Hussein told me. "But he was a roughneck, he was a fighter. That's what he was known for."

Hussein recalled that Berjawi was associated with a Muslim gang in north London that was embroiled in fights with rival Irish youths. But he saw no sign that Berjawi was involved in anything other than unruly teenage behavior. On one occasion Berjawi and a group of his friends were given the opportunity to go on a vacation overseas, funded by the community organization Hussein worked for. Hussein recalls that the group chose a holiday resort in Benidorm, on the east coast of Spain, where they were thrown out of a hotel for being too

Church Street Market near Edgware Road, northwest London. September 29, 2015.

raucous. "They got up to such craziness, smashed up a hotel room," he said. "I never saw him drinking, but obviously he got up to – he liked all the stuff that young guys like, partying and stuff like that."

The period between 2003 and 2006 appears to have been a crucial and formative time in Berjawi's life, when he transitioned from partying in Spain and playing soccer in London parks to joining up with al Qaeda–affiliated militants in Somalia. According to the Pentagon case study, in 2006 Berjawi left London for a short period and attended a training camp called "Bayt al-Jinn," where he received explosives training. He then "returned to the U.K. and provided financial support to [al Qaeda] allied elements in East Africa." The case study does not specify the location of the Bayt al-Jinn camp. However, a previously secret detainee report on a Kenyan terror suspect held at Guantánamo, published by WikiLeaks in 2011, mentions a "Bayt Jinn House" in Mogadishu that was allegedly frequented by international al Qaeda operatives in the region.[3] The Guantánamo report also states that a group known as the "London boys," of which Berjawi was a member, attended a training camp in Mogadishu in the fall of 2006.

The U.S. government accounts are corroborated by a martyrdom biography of Berjawi published on jihadi Internet forums after his

Abbey Road in St. John's Wood, London. September 29, 2015.

death, which states that he "joined with the Mujahideen in Somalia during the time of the Islamic Courts Union," referring to a coalition of Sharia courts that gained control of large parts of Somalia in 2006.[4] There he attended his "first training," according to the biography, then returned to the U.K., where he took responsibility for "the collection of funds and its delivery."

After returning from Somalia in 2006, Berjawi does not appear to have had any direct contact with British police or security agencies. Despite his apparent instruction at an al Qaeda–affiliated camp, he was not arrested on his way back to England, suggesting that intelligence collected by the United States about his whereabouts might not have been immediately shared with British agencies. Lynne Arnold, a spokesperson for London's Metropolitan Police, declined to answer questions for this story, saying she was "not able to discuss" why Berjawi was not arrested or whether U.S. authorities had shared any information about him.

According to the interview conducted by Cage, which campaigns on behalf of terrorism suspects who are denied legal rights, Berjawi did not begin to notice that British authorities were interested in him until about 2007. That year counterterrorism forces in Nairobi detained two of Berjawi's friends from London, who had fled Somalia after war broke out with Ethiopia. The pair were later released without charge. Upon their return to London, the men told Berjawi that during their detention in Kenya, British agents had questioned them and shown them his photograph. "That's when I realized myself I was starting to be followed," Berjawi told Cage. "I would see someone – the same person – following me, wherever I was. The same car – I actually even memorized the number plate."

Berjawi's suspicions appear to have been further confirmed between 2007 and 2008. During a trip to Lebanon he was stopped at a Lebanese airport and asked why he had traveled to the country. He told the authorities he was visiting family and gave them a phone number for his uncle, and eventually they let him through. He was interrogated again on his way out of Lebanon but arrived back in London without any problems. A few days after his return, however, Berjawi called his uncle and learned that he had been approached by Lebanese counterterrorism agents, who had been asking questions about Berjawi.

By early 2009 Berjawi was working in London with his stepfather as a plumber and air-conditioning engineer. He had gotten married and had a baby girl, and his wife was pregnant with another child, this time a boy. But Berjawi was still on the radar of security agencies, and he was about to experience his first serious interrogation. With a childhood friend named Mohamed Sakr, Berjawi arranged a trip to Kenya. According to his account, he wanted to go on a wildlife safari, but counterterrorism officials in Kenya suspected otherwise. When he arrived at the Mombasa airport, Berjawi was stopped and questioned. He was permitted into the country but noticed a man of Somali origin following him everywhere, whom he suspected was some sort of spy.

"Wherever I go to eat, whatever safari park we go to, he's always there on his phone," Berjawi told Cage. "When I stop, he stops; when I walk, he walks."

After a few days in Mombasa, Berjawi and Sakr traveled to Nairobi, perhaps in an effort to avoid the man they believed was tailing them. The pair stayed at the family home of Naji Mansour, an American citizen living in Nyari, an affluent Nairobi neighborhood located near the United Nations Africa headquarters. Mansour, who was thirty-two at the time, lived with his wife and two children in a large house in a compound with its own gym, game room, and garden. The main part of the house had four bedrooms, and there were two additional bedrooms in a separate wing that the family kept for guests.

Recalling how he first came into contact with Berjawi and Sakr, Mansour told me that he put them up as a favor to a friend named Mohamed, whom he had met in Dubai while working briefly for a tech company there that provided information security services. According to Mansour, Berjawi and Sakr claimed they had traveled to Kenya to research a substance known as miraa, or khat, an amphetamine-like stimulant grown and consumed in the Horn of Africa and the Arabian Peninsula. Mansour's first impression of the pair was that they were "regular Joes." They lounged around the house, watched movies, played games with his children, and occasionally prayed. "But they didn't seem like hardcore, staunch Muslims," Mansour said.

At first Berjawi and Sakr said they would need a place to stay for only a few days. But a few days soon turned into a week. When Mansour asked about their plans, he was told they were waiting for some money to be sent to them before they moved on. "I didn't feel like they were a threat in any way, even when they overstayed," said Mansour. "The only strange thing that I noticed from them the whole time is that it seemed like they weren't trying to go out; they weren't trying to leave the house."

One day, about two weeks into their stay, in February 2009, Kenyan antiterror police surrounded Mansour's Nairobi house. Berjawi was playing pool in the game room when he heard a loud series of knocks at the door. He peered through a curtain and saw heavily armed Kenyan officers, a helicopter flying above, and lots of cars. The police then stormed the property, told Berjawi to get on the floor, and pointed a gun at his head while he was searched. Berjawi told Cage that he and Sakr were handcuffed, taken to the antiterror police headquarters, and placed in separate cells. Berjawi described his cell as a "black hole" with "no pillows, no light, nothing," and said that when he asked for food a guard told him that he had to drink his own urine.

Later that day, Berjawi said, he was taken from his cell through a long dark corridor to a private room. He was dazzled by the bright lights when the door opened, and when his eyes regained focus he could see about five men, dressed smartly in suits. "They looked like professional people, y'know, they didn't look like they belonged there," Berjawi later recalled. "You could tell the difference between them and the guards that were working there. With the guards you can smell the sweat on them, and some of them were even drunk." Berjawi said the men accused him of being an al Qaeda suicide bomber who had come to Kenya as part of a plot to attack the Israeli Embassy and an Israeli-owned supermarket. He denied the allegations and requested a lawyer. "My friend, this is Africa," he recalled being told. "In Africa, the only thing we can give you is black magic."

For four days Berjawi and Sakr were held in custody and repeatedly interrogated. According to Berjawi, when he was eventually given some food, a porridge-like dish called ugali, the guards had sprinkled it with cigarette ash. He claimed they also asked him if he was gay and insinuated that they were going to send in a man who would rape him. Toward the end of the ordeal Berjawi said that both he and Sakr endured several mock executions. "They just threw us out [of] the car in the forest, and we heard 'tchck-tchk' – you know, the noise was there, and then I'd feel a gun to the back of my head, like that, but . . . nothing. Then they'd just all laugh, pick us back up, throw us back into the car, then they'd drive again. They did this twice or three times." (Kenya's National Police Service, the authority responsible for law enforcement in the country, did not respond to requests for comment.)

According to Berjawi, there were no British agents present during his interrogations in Kenya. He did believe, however, that British government operatives were feeding questions to the Kenyans, who seemed to know many highly specific details about his life in London, such as his daughter's name, where he played soccer, the names of his friends, and which mosque he attended. On the final day of his detention a woman Berjawi said was from the British Embassy visited him, asked how he was doing, and handed him some forms to fill out. Shortly afterward he was released. Together with Sakr, Berjawi was flown back to London accompanied by four Kenyan agents.

When the plane touched down, an announcement came over the speakers instructing all passengers to remain in their seats. A large group of "big white built men came on the plane [in] suits," Berjawi later recounted. "One of them directly looked at me and smiled, and he called me, 'Bilal, would you like to stand up?'" The men ushered Berjawi and Sakr off the plane, at which point the friends were separated. The men told Berjawi they were from the British domestic security agency, MI5.

Over a period of about ten hours the agents interrogated him about his visit to Kenya. They warned him that he was not allowed to decline to answer their questions, suggesting he was detained under the Terrorism Act, which makes it a criminal offense to respond "No comment."[5] The British agents snapped photographs of Berjawi and took his fingerprints. He recalled that they were apologetic, telling him, "We have to do this." But he was left feeling aggrieved; after interrogating him, the agents took his money and shoes, handed him his clothes in a garbage bag, and left him alone in the airport, barefoot, without any means to return to his home in northwest London.

A few weeks later Berjawi called his uncle in Lebanon, who described receiving another visit from counterterrorism agents. This time the agents informed him that Berjawi was "involved in al Qaeda," based on "information from Britain." They emphasized that his nephew shouldn't return to Lebanon or there would be problems.

Meanwhile Berjawi began to suspect that he was being followed each time he set foot outside his London home. On one occasion, shortly after he returned from Kenya, he went out to the supermarket and noticed two men who he believed were tailing him. On the street he bumped into an old friend and stopped for a quick conversation. Berjawi said that the two men subsequently approached his friend, who was taken away in a car to a nearby police station and interrogated. The increased scrutiny appears to have agitated and unsettled Berjawi, though he still had not been arrested in the U.K. or charged with any crimes.

In April 2009 he approached Cage to complain that he was being "harassed" by security services, according to the transcript of his interview. "I don't want to be harassed, followed – I feel intimidated, I've got a lot of side effects, you know," Berjawi told the advocacy group. "My friends have been scared away from me because they've been approached. I feel isolated. . . . It's becoming a bit too much."

Within six months, in October 2009, Berjawi and his friend Sakr were back in Somalia. A year later, in September 2010, the British government revoked the passports of both men under the British Nationality Act, severing its legal obligations to uphold their rights as citizens, a move that may have paved the way for their assassination.

Berjawi wanted to appeal the decision to revoke his passport and in October 2010 sent an email to a contact at Cage asking the organization to instruct his lawyer, Saghir Hussain, to represent him in the case. Hussain told me that there were difficulties filing the appeal, primarily because of security concerns about talking over the phone to Berjawi in Somalia.

> "I said to his family, 'Look, I can't guarantee that while he's communicating with us he won't be droned and killed,'" Hussain recalled. "That's why it was decided that it was too risky for us to carry on." As it turned out, Hussain's concerns were well-founded.

According to a slide titled "FFF Timeline: Objective Peckham Case Study" in the ISR study since 2006 the covert Joint Special Operations Command unit TF 48-4 had been keeping close tabs on Berjawi's movements. He had been featured on a so-called baseball card used by the U.S. government to encapsulate information about candidates for assassination, and had thus entered the kill/capture process known as "find, fix, finish." By December 2009, the document alleges, Berjawi was helping to "facilitate money, equipment, and fighters" through the U.K. to Somalia. Throughout 2010 the U.S. government collected intelligence on him from intercepted communications, and before long operators pinpointed his location. On June 23, 2011, Berjawi was tracked to an area near Kismayo, a port city some 250 miles from Mogadishu. The special operations unit launched a missile strike, according to the document, but it was unsuccessful due to a malfunction and other problems related to "approval authorities."

Church Street Market near Edgware Road, London. September 29, 2015.

Although Berjawi was not killed, he may have been wounded in the attack or in another carried out around the same time. On June 24 the Associated Press reported a missile strike late the previous day on a convoy of al Qaeda–linked militants near Kismayo, which injured two or three of the fighters.[6] Two weeks later Somali media reported that Berjawi, a "senior officer" with al Qaeda, was believed to have been injured in an attack and had traveled to Kenya for medical treatment.[7] It was not until the following year that U.S. forces again identified Berjawi's location.

According to the "FFF Timeline: Objective Peckham Case Study," on January 21, 2012, Berjawi's white SUV was observed at 3:59 a.m., presumably by drone, and his movements were tracked over several hours in an area a few miles northwest of Mogadishu, between the towns of Afgooye and Ceelasha. The timeline describes an "adult with heavy strides and slight limp (OBJ PECKHAM)" at 5:02 a.m. Three hours later, at 8:11, a "vehicle follow begins." At 10:39 surveillance equipment logged a "Full Register/Match" of a cell phone in the target area, meaning the unique identifying codes of a SIM card and handset associated with Berjawi had been confirmed by the special operations unit. Twenty-four minutes later, at 11:03, Bilal el-Berjawi, otherwise known as Objective Peckham, "was eliminated via kinetic strike," the entire front half of his vehicle mangled by the explosion. The timeline of the strike, oddly, shows another match with the cell phone at 11:31 a.m. The drone continued "to monitor the scene."

The following day a spokesperson for al Shabaab calling himself Sheikh Ali Mohamud Rage confirmed the death of Berjawi, whom he described as a senior al Qaeda commander in Somalia.[8] Rage said that Berjawi had been killed by a U.S. drone, and he vowed revenge for the killing. He added, "We take his death as congratulation, thanks to Allah. . . . His martyrdom dream has just become true."

As news spread of Berjawi's death, it fueled paranoia within elements of al Qaeda in Somalia. Seven months earlier, al Qaeda's chief in East Africa, Fazul Abdullah Mohammed, had also been killed.[9] Berjawi was said to have been close to Mohammed, and perhaps was his successor, so when he too died in a sudden attack there were suspicions that al Shabaab was carrying out an internal coup. Indeed some news reports out of Kenya initially suggested the attack on Berjawi was an "inside job" and that he had been assassinated due to a power struggle.[10] Subsequently one Somali outlet reported that at least one hundred foreign al Qaeda fighters in Somalia had fled the country, partly due to leadership squabbles.[11] "It is true that those brothers left us and went to Yemen due to some minor internal misunderstandings amongst ourselves," an al Shabaab spokesperson was quoted as saying at the time. "This started when we lost our brother, Bilal el-Berjawi, on January 21."

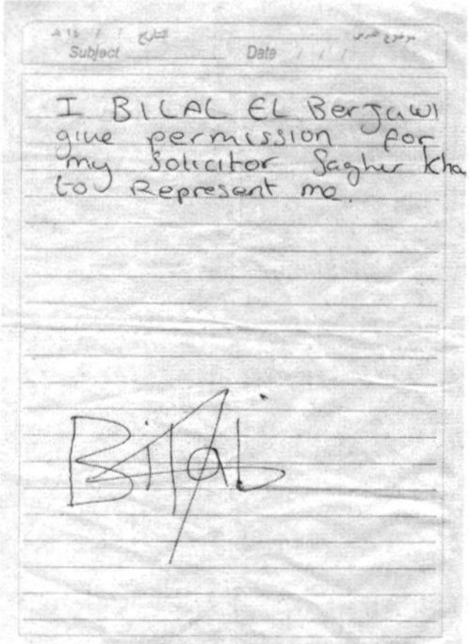

Subject Date

I BILAL EL Berjawi give permission for my Solicitor Saghir Kha to Represent me.

Bilal

A note written by Berjawi in October 2010, after his U.K. citizenship was revoked. No appeal was filed.

Once it became apparent that Berjawi had in fact been killed in a U.S. drone strike, the groups appear to have settled their differences and strengthened their alliance. Three weeks after Berjawi's death, the leaders of al Qaeda and al Shabaab appeared in a video together. Al Shabaab pledged its allegiance to al Qaeda and vowed that it would "march with you as loyal soldiers."[12]

Shortly before Berjawi was killed, his wife had given birth to a boy in London. She is believed to have spent time with Berjawi in Somalia but had returned to London in 2011.

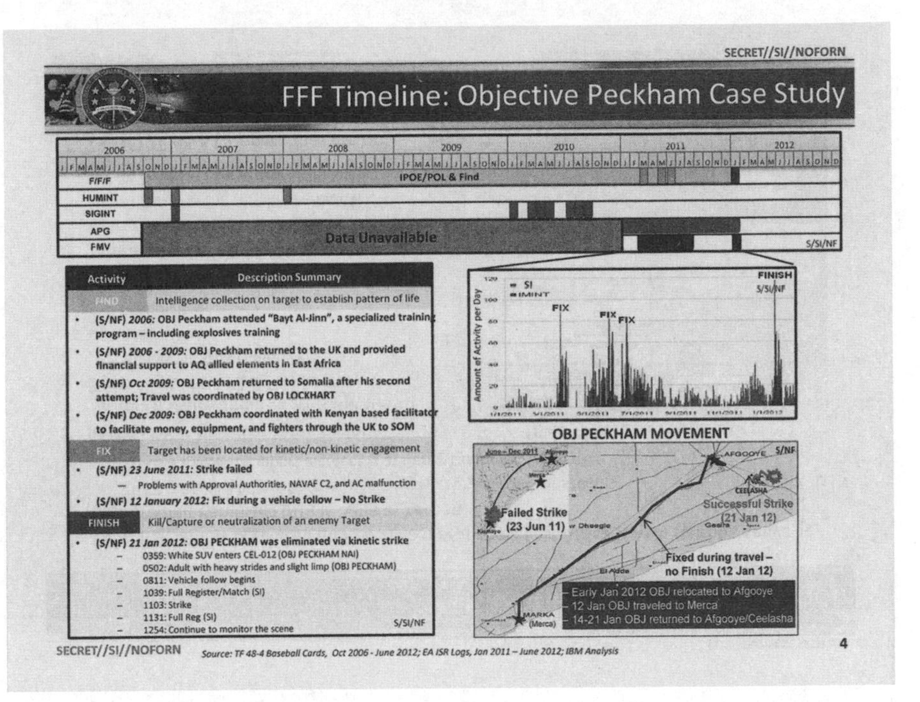

A secret case study details the surveillance and assassination of "Objective Peckham," Bilal el-Berjawi, in January 2012.

Upon hearing about the birth of his third child, Berjawi reportedly phoned his wife while she was in the hospital, hours before he was killed. Relatives speculated that it was this phone call that had exposed him as a target for the drone strike. That seems unlikely, however. According to the Pentagon timeline, Berjawi's location had already been established by the covert special operations unit nine days prior to the lethal attack. Cell phone surveillance helped pinpoint him on the day he died, but it is unclear whether the phone in question belonged to him or had been covertly placed in his vehicle by someone else to aid the strike.

Six months after Berjawi's death, in July 2012, al Shabaab publicly executed three men accused of helping British and American spy agencies kill Berjawi.[13] In a propaganda video the alleged informants confessed to having hidden a cell phone in Berjawi's vehicle so that he could be tracked and bombed. One of the accused informants, Isaac Omar Hassan, said a man working with the CIA in Mogadishu handed him a Nokia X2 cell phone and an envelope containing $4,000 cash. He was asked to place the phone in Berjawi's

Lisson Grove in northwest London, September 29, 2015.

vehicle and make sure it was turned on when requested, which he said he did on the day Berjawi was targeted.

In February 2012, about a month after Berjawi's death, his childhood friend Mohamed Sakr was also killed in a reported U.S. drone strike in Somalia.

The revocation of Berjawi's and Sakr's passports prior to their deaths by U.S. drone strike has raised questions about whether the British government was secretly complicit in their assassination.[14] Ben Stack, a spokesperson for the U.K. Home Office, declined to comment for this story when asked whether the passports were revoked as part of a coordinated sequence of events that culminated in deadly attacks by U.S. special operations forces. "We don't routinely comment on security matters," he said.

Kat Craig, the lawyer with the London-based human rights group Reprieve, told me that she believed there was "mounting evidence" that the British government has used "citizenship-stripping" as a tactic to remove legal obstacles to killing people suspected of being affiliated with terrorist groups. "If the U.K. government had any role in these men's deaths – including revocation of their citizenship to facilitate extrajudicial killings – then the public has a right to know," Craig said. "Our government cannot be involved in secret executions. If people are accused of wrongdoing they should be brought before a court and tried. That is what it means to live in a democracy that adheres to the rule of law."

Since 2006 the British government has reportedly deprived at least twenty-seven people of their citizenship on national security grounds, deeming their presence "not conducive to the public good."[15] The power to revoke a person's citizenship rests solely with a government minister, though the decision can be challenged through a controversial immigration court. When cases are brought on national security grounds, they are routinely based on secret evidence, meaning the accusations against individuals are withheld from them and their lawyers. "The net effect of the practice," according to Craig, is "not only to remove judicial oversight from a possible life-and-death decision, but also to close the doors of the court on anyone who seeks to expose some of the gravest abuses being committed by Western governments."

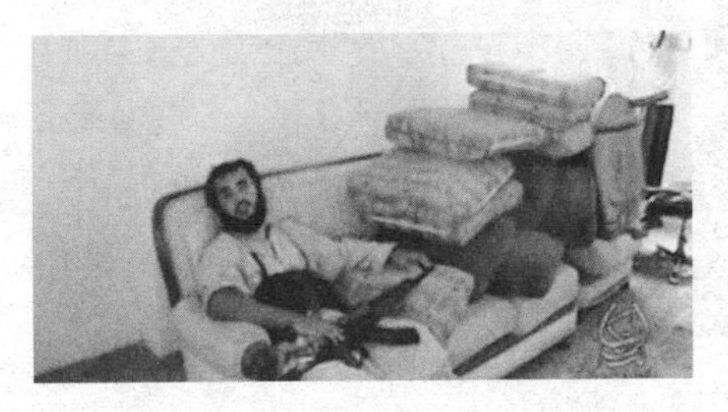

Bilal el-Berjawi holding an AK-47 rifle in a martyrdom video produced after his death by al Shabaab's media wing.

There have reportedly been at least ten British citizens killed in drone attacks as part of a covert campaign that, between 2008 and 2015, has gradually expanded from Pakistan to Somalia and now to Syria.[16] In late August 2015 an Islamic State computer hacker named Junaid Hussain, a former resident of Birmingham, England, was assassinated on the outskirts of Raqqa, Syria, by a U.S. drone strike.[17] Several days earlier, in another attack near Raqqa, the U.K. government deployed its own drones for the first time to target British citizens, killing Islamic State recruits Ruhul Amin and Reyaad Khan while they were traveling together in a car.[18] It remains unclear whether, like Berjawi and Sakr, these targets had their British passports revoked before they were killed. Stack, the Home Office spokesperson, would not discuss the citizenship status of Hussain, Amin, Khan, or other Britons killed by drones. "We don't talk about individual cases, and also we don't comment on matters of national security," he told me.

Regent's Park Mosque, northwest London, September 29, 2015.

Around the community in which Berjawi grew up the reverberations of his life and death continue to be felt. News reports have featured his name as a one-time associate of Mohammed Emwazi, better known as the masked Islamic State executioner nicknamed "Jihadi John." Emwazi lived near Berjawi in northwest London, and a source familiar with his circle of friends told me that the pair had attended the same school. Emwazi was a few years younger than Berjawi and "looked up" to him, according to the source, who asked not to be named. Several of Berjawi's former friends still live and work in London but have distanced themselves from the controversy surrounding him. One of Berjawi's closest former friends now works as a bus driver; another of his peers has since become an imam. Many, including Berjawi's family members and neighbors, are reluctant to talk about him publicly.

On the quiet tree-lined street in London where Berjawi spent his youth, cars come and go, and a new generation of children laughs and plays out on the sidewalk. At Berjawi's former apartment, where some members of his family still live, is a creased Arabic poster pinned to the door with a message for visitors. "Whoever believes in God and the Judgment Day," it reads, "let him speak up, or remain silent."

For many years lawyers and human rights advocates have wondered about the chain of command in cases of non-battlefield assassinations. Who authorizes them? Do they fall within the 2001 Authorization for Use of Military Force (AUMF), or through some other authority?

The secret documents are not comprehensive on this point, but they suggest a linear chain—all the way up to the president of the United States (POTUS).

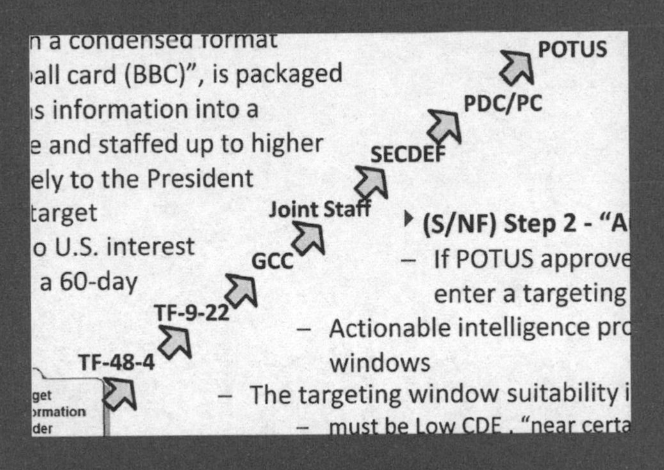

MANHUNTING IN THE HINDU KUSH

RYAN DEVEREAUX

From 2011 to 2013 the most elite forces in the U.S. military, supported by the CIA and other elements of the intelligence community, set out to destroy the Taliban and al Qaeda forces that remained hidden among the soaring peaks and plunging valleys of the Hindu Kush, along Afghanistan's northeastern border with Pakistan. Dubbed Operation Haymaker, the campaign has been described as a potential model for the future of American warfare: special operations units, partnered with embedded intelligence elements running a network of informants, pinpointing members of violent organizations, then drawing up plans to eliminate those targets from the battlefield, either by capturing or killing them.

Classified intelligence community documents published by *The Intercept* on October 15, 2015, detail the purpose and achievements of the Haymaker campaign and indicate that the American forces involved in the operations had, at least on paper, all of the components they needed to succeed. After more than a decade of war in Afghanistan, a robust network of intelligence sources, including informants on the ground, had been established in parts of the historically rebellious, geographically imposing provinces of Kunar and Nuristan. The operators leading the campaign included some of the most highly trained military units at the Obama administration's disposal, and they were supported by the world's most powerful electronic surveillance agencies, equipped with technology that allowed for unmatched tracking of wanted individuals.

Despite all these advantages, the military's own analysis demonstrates that the Haymaker campaign was in many respects a failure. The vast majority of those killed in airstrikes were not the direct targets. Nor did the campaign succeed in significantly degrading al Qaeda's operations in the region. When contacted with a series of questions regarding the Haymaker missions, the U.S. Special Operations Command in Afghanistan declined to comment on the grounds that the campaign, though now finished, remains classified.

The secret documents obtained by *The Intercept* include detailed slides pertaining to Haymaker and other operations in the restive border regions of Afghanistan, including images, names, and affiliations of alleged militants killed or captured as a result of the missions; examples of the intelligence submitted to trigger lethal operations; and a "story board" of a completed drone strike. The targets identified in the slides as killed or detained represent a range of militant groups, including alleged members of the Taliban and al Qaeda, but also local forces with no international terrorism ambitions, groups that took up arms against the United States after American airstrikes brought the war to their doorstep.

One slide charts mission statistics from September 2011 to September 2012 for Task Force 3-10, which was responsible for special operations across Afghanistan, breaking down in rare detail the more than two thousand missions conducted by elite U.S. forces in the country over the course of a year.

Together the materials offer an unprecedented glimpse into the kind of killing that has come to define the war on terror, underscoring the inherent limitations, and human cost, of those operations. With the Obama administration publicly committed to continuing campaigns like Haymaker – special operations missions focused on hunting down specific individuals not only in Afghanistan but in nations around the world – the documents raise profound questions about the legacy of the longest foreign war in American history and its influence on conflicts to come.

The frequency with which "targeted killing" operations hit unnamed bystanders is among the more striking takeaways from the Haymaker slides. The documents show that during a five-month stretch of the campaign, nearly nine out of ten people who died in airstrikes were not the Americans' direct targets. By February 2013 Haymaker airstrikes had resulted in no more than thirty-five "jackpots," used to signal the neutralization of a specific targeted individual, while more than two hundred people were declared EKIA, "enemy killed in action."

A village security force commander and coalition special operations forces identify insurgent fighting positions during a daylong firefight in Nuristan Province, Afghanistan, April 12, 2012.

In the complex world of remote killing in remote locations, labeling the dead as "enemies" until proven otherwise is commonplace, said an intelligence community source with experience working on high-value targeting missions in Afghanistan, who provided the documents on the Haymaker campaign. The process often depends on assumptions or best guesses in provinces like Kunar and Nuristan, the source said, particularly if the dead include "military-age males"

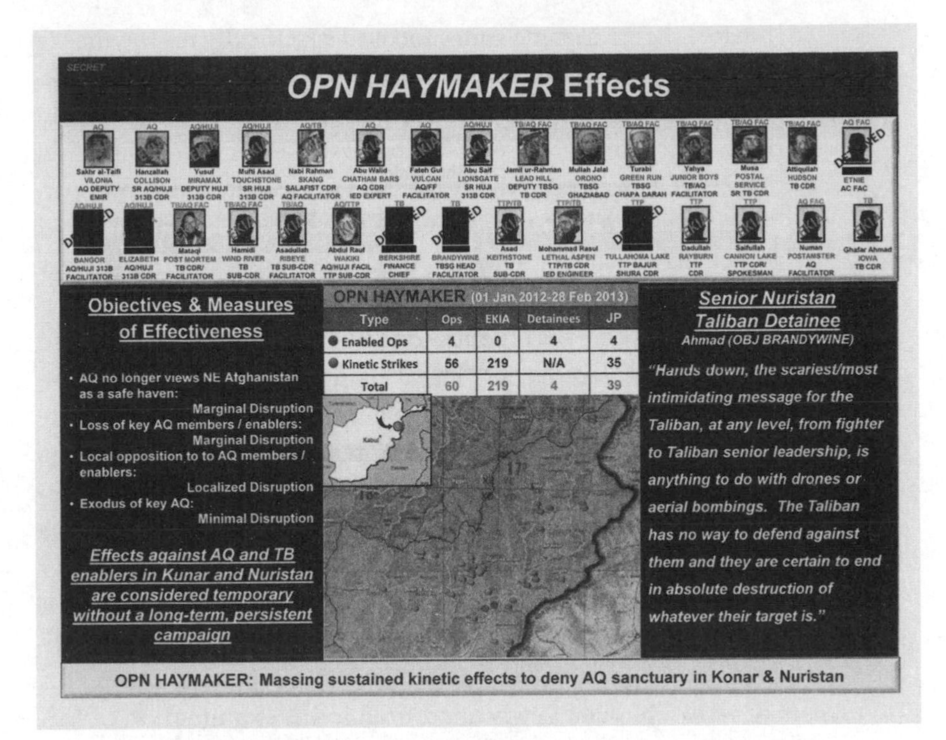

A summary of operations conducted from January 2012 through February 2013 as part of Operation Haymaker, a U.S. military campaign in northeastern Afghanistan aimed at rooting out elements of al Qaeda and the Taliban.

(MAMs, in military parlance). "If there is no evidence that proves a person killed in a strike was either not a MAM, or was a MAM but not an unlawful enemy combatant, then there is no question," he said. "They label them EKIA." In the case of airstrikes in a campaign like Haymaker, the source added, missiles could be fired from a variety of aircraft. "But nine times out of ten it's a drone strike."

The source is deeply suspicious of those airstrikes – the ones ostensibly based on hard evidence and intended to kill specific individuals – which end up taking numerous lives. Certainty about the death of a direct target often requires more than simply waiting for the smoke to clear. Confirming that a chosen target was indeed killed can include days of monitoring signals intelligence and communication with sources on the ground, none of which is perfect 100 percent of the time. Firing a missile at a target in a group of people, the source said, requires "an even greater leap of faith" – a leap that he believes often treats physical proximity as evidence.

The documents include slides focused specifically on Haymaker operations from January 2012 to February 2013, distinguishing between raids, described as "enabled" and "combined" operations, and airstrikes, which are described as "kinetic strikes." In both raids and airstrikes, the source said, the target was always an individual person: "Every mission that's triggered begins as an objective to find one person for whatever reason," the source said, adding, "Every jackpot is one person off the list."

According to the documents, raids performed on the ground during Haymaker were far less lethal than airstrikes and led to the capture of scores of individuals. Research by Larry Lewis, formerly a principal research scientist at the Center for Naval Analyses, supports that conclusion. Lewis spent years studying U.S. operations in Afghanistan, including raids, airstrikes, and jackpots, all with an eye to understanding why civilian casualties happen and how to better prevent them. His contract work for the U.S. military, much of it classified, included a focus on civilian casualties and informed tactical directives issued by the top generals guiding the war. During his years of research, what Lewis uncovered in his examination of U.S. airstrikes, particularly those delivered by machines thought to be the most precise in the Pentagon's arsenal, was dramatic. He found that drone strikes in Afghanistan were ten times more likely to kill civilians than conventional aircraft.[1]

"We assume that they're surgical, but they're not," Lewis said in an interview. "Certainly in Afghanistan, in the time frame I looked at, the rate of civilian casualties was significantly higher for unmanned vehicles than it was for manned aircraft airstrikes. And that was a lot higher than raids."

The limited point of view of the drone's camera, what Lewis describes as the "soda straw effect," together with the globally dispersed operational network that supports drone strikes, can lead to mistakes, he argues, including the loss of innocent lives. The materials obtained by *The Intercept* make just one explicit mention of civilian casualties, in the Task Force 3-10 mission statistics from September 2011 through September 2012. The document reveals that the United States conducted more than 1,800 "night ops" at a time when Afghanistan's president Hamid Karzai was calling for an end to American involvement in controversial night raids.[2] Of those operations – which resulted in 1,239 targets captured or killed and

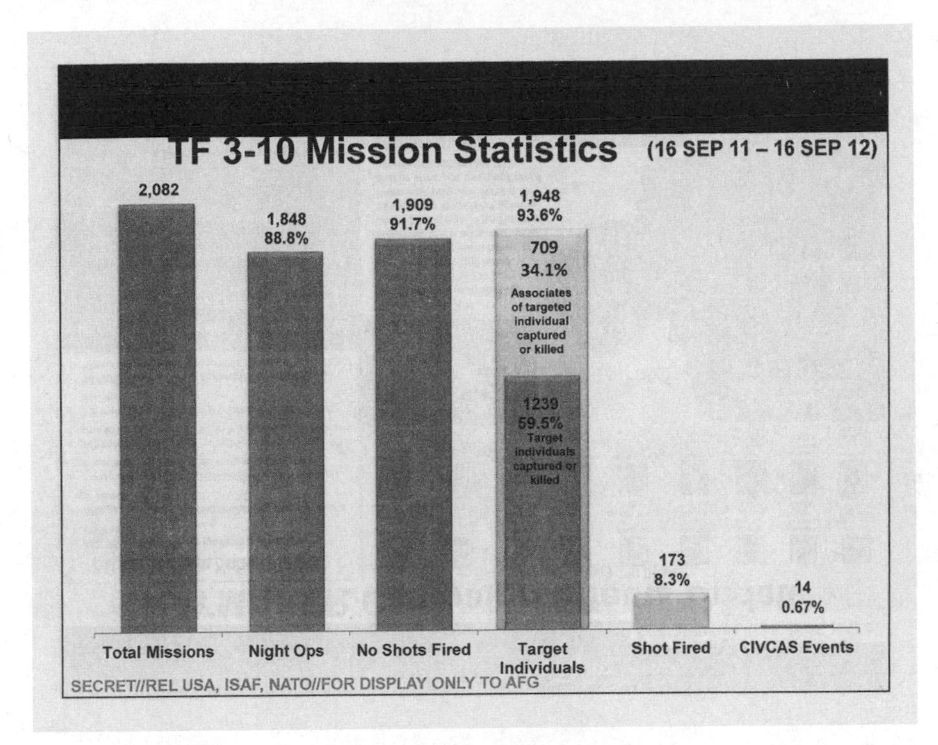

A breakdown of 2011-12 mission statistics for Task Force 3-10, a U.S. special operations task force responsible for missions in Afghanistan at that time.

709 "associates" of targets captured or killed – the military reported "shots fired" in fewer than 9 percent of its missions, with a total of fourteen civilian casualty "events" for the year.

"The fourteen civilian casualties is highly suspect," said the source, who reviewed after-action reports on raids and other operations in Afghanistan. "I know the actual number is much higher," he added. "But they make the numbers themselves so they can get away with writing off most of the kills as legitimate."

The Haymaker documents reveal little about whether or not the deaths reflected in the materials were "legitimate." They do, however, offer an illustrative window into how the killing has been done in the past – and how it may be done in the future.

The request was unambiguous. Dated October 30, 2012, and stamped with the seal of U.S. Central Command, the document is titled, "Request for Kinetic Strike Approval." The "desired results" listed at the top of the document include just three words: "Kill Qari Munib."

U.S. special operations personnel prepare to board a UH-60 Black Hawk helicopter during a mission in Kunar Province, Afghanistan, February 25, 2012.

Munib, whose objective code name was "Lethal Burwyn," was described as a Taliban subcommander operating in the Pech district of Kunar Province. He allegedly exercised command and control over a specific portion of the organization, was responsible for numerous attacks on both coalition and Afghan security forces, and communicated with Taliban officials in Pakistan. Specifically, the request reports, Munib had been implicated in recent plots to carry out improvised explosives attacks.

The Americans considered the consequences of taking Munib's life, including media coverage, possible political fallout, and potential "population blowback." It was determined that negative repercussions were "unlikely" in all three categories and that Munib's death would "decrease attacks" on coalition and Afghan forces. Going through with the operation, the request asserts, would require a signals intelligence "correlation," followed by a full-motion video lock, visual identification within twenty-four hours of the strike, and a "low" probability of collateral damage. Two maps are featured in the document intended to seal Munib's fate, one of which includes coordinates of his last known location. In the bottom right-hand corner of the document is a bar, numbered one to ten and fading in color

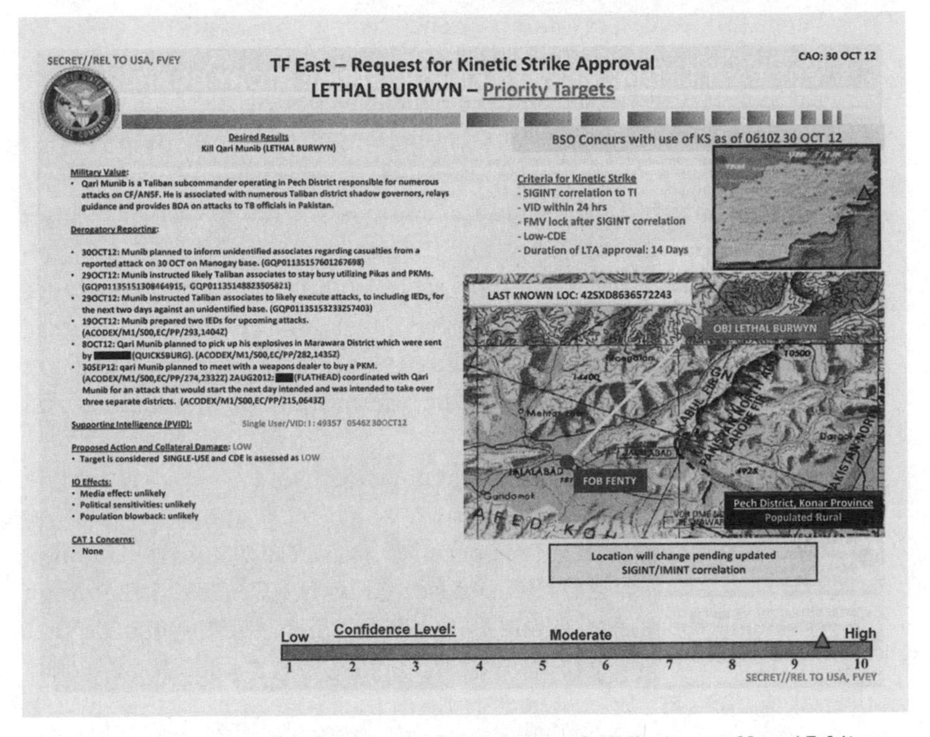

A 2012 U.S. military request for an airstrike targeting Qari Munib, an alleged Taliban subcommander operating in northeastern Afghanistan.

from red to green. It is titled "Confidence Level." A red triangle sits between the numbers 9 and 10.

Less than a week after the briefing was completed, the kill mission was under way. Signals intelligence had been picked up from a compound where Munib was known to sleep, according to a storyboard detailing the operation. Images relayed from the scene revealed the presence of five military-age males in the area. Floating above the site, an MQ-9 Reaper drone, known as "Skyraider," captured the image of a man dressed in a drab, flowing robe, with a white cap on his head, casting a long shadow in the dirt. According to the storyboard, the image was "correlated" to signals intelligence linked to Munib.

Skyraider loitered above as the man, joined by two others, walked up a ridgeline before heading back into the compound. The figure again emerged from the building. The drone's camera registered a positive identification. Skyraider "engaged." A screen grab from the scene shows a cloud of smoke where the individual had been. Task force personnel watched as a group of people collected the target's remains. "RESULTS: JP – Pending EKIA, 1 x TOTAL EKIA," the storyboard reports. An enemy had been killed in action. Confirmation that he was indeed Munib, the jackpot, or JP, was still pending.

On Friday, November 9, 2012, NATO issued one of its standard updates on missions around the country, including two short lines about an operation carried out the day before in eastern Afghanistan: "An Afghan and coalition security force killed Taliban leader Qari Munib during a security operation in Kunar province Thursday. Qari Munib was responsible for directing attacks against Afghan and coalition forces and coordinating the movement of weapons and ammunition for the attacks."[3]

However, the announcement appeared five days after the drone strike described in the storyboard, which made no mention of Afghan forces involved in the operation. Was Munib killed in a unilateral U.S. drone strike, later obfuscated by NATO? Or did the drone strike fail to jackpot, resulting in a subsequent joint operation that succeeded in eliminating him? If so, who was it that Skyraider engaged that day? Whose body parts did the American analysts watch the first responders collect?

Those questions remain unanswered. A more fundamental question suggests itself, however: How did the most powerful military in history come to devote its elite forces and advanced technology to the hunt for a man like Qari Munib, a midlevel Taliban figure in a remote corner of the planet, half a world away from the White House and ground zero in Manhattan, more than eleven years after the September 11 attacks?

When the Americans set out to kill Qari Munib with a drone in 2012, an intelligence document purporting to lay out his bona fides as a target listed local insurgency figures alongside regional actors. In a graphic titled "Link Analysis," Munib's name appears under a generic cartoon of an Afghan male, surrounded by six other headshots. Half of them are described as Salafists, a conservative faction that has existed in Kunar for decades and, for a period, resisted Taliban presence in the province.

One of the Salafists pictured is Haji Matin, a timber trader from the Korengal Valley. In the early years of the war one of Matin's business rivals wrongly fingered him to the Americans as a militant. U.S.

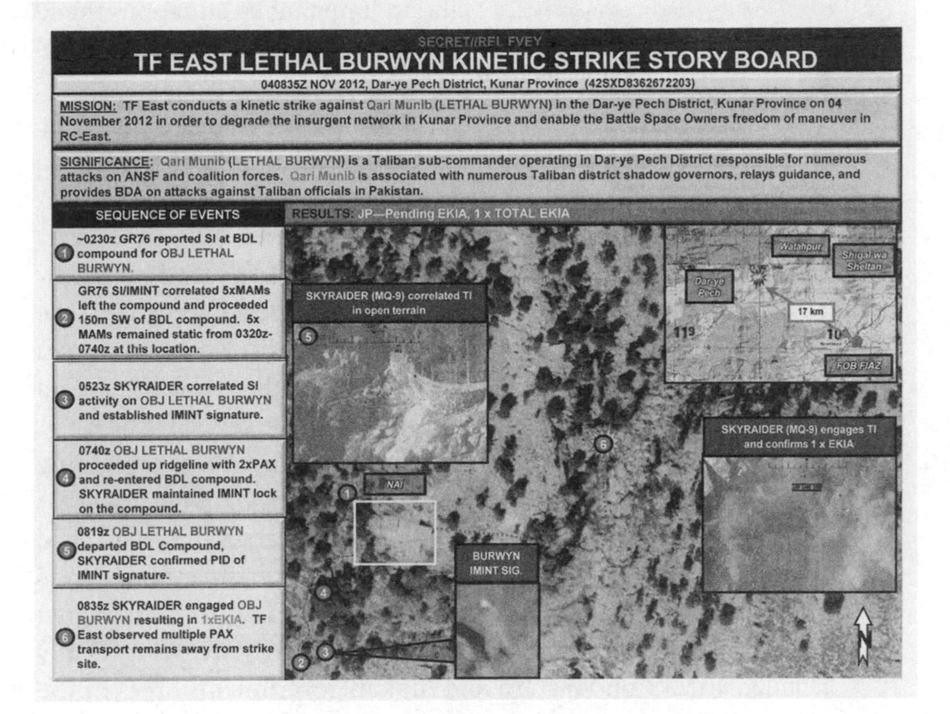

A storyboard detailing a 2012 U.S. drone strike targeting Qari Munib, an alleged Taliban subcommander.

forces responded by bombing Matin's home. While Matin survived, several members of his family were killed.[4] The Americans then appropriated one of his lumberyards as an outpost, thus turning one of the most powerful men in the area into a formidable insurgent leader. The transformation of men like Matin and the Salafists, once locally minded power brokers, into anti-U.S. fighters was hardly unique.

"When viewed from absolutely the wrong metric, the Americans were very successful at hunting people," said Matt Trevithick, a researcher who in 2014 made more than a dozen unembedded trips to some of Kunar's most remote areas in an effort to understand the province, and American actions there, through the eyes of its residents.[5] The problem, he said, is that savvy, opportunistic strongmen maneuvered to draw U.S. forces into local conflicts, a dynamic that played out again and again throughout the war. "We knew nothing about who we were shooting at – specifically in Kunar," Trevithick said. He understands the frustration of conventional U.S. forces who were dropped in places like Kunar. "I don't blame them. They're put

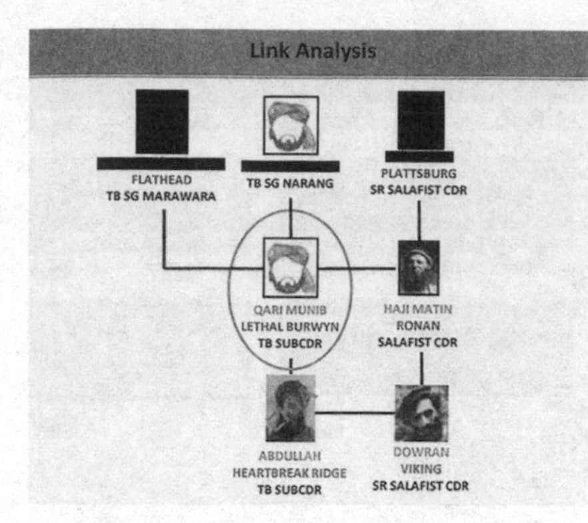

A "link analysis" purporting to detail alleged Taliban subcommander Qari Munib's ties to other militant figures.

in an impossible situation themselves. But what happens is everyone starts looking like the enemy. And that means you start shooting. And that means people actually do become the enemy."

In September 2010, nine years after the terrorist attacks in New York City and Washington, D.C., the U.S. military and coalition forces were working their way through a list of 744 people slated for death or capture in Afghanistan. According to the so-called Joint Prioritized Effects List (JPEL), provided by NSA whistleblower Edward Snowden, Kunar, with forty-four targets, had the third-highest total in the country. Few of the Kunar targets represented core al Qaeda–linked figures, and many were associated with local groups like the Salafists, whose listed offenses typically included attacks on Western and Afghan government forces in the province. The targets on the 2010 list were prioritized with rankings of 1 to 4, in terms of their significance, with 1 being the most significant. In Kunar a single target rose to the level of priority 1, while more than 80 percent were designated priority 3. Seven of the JPEL targets appear in the Haymaker slides, though just three had been linked to al Qaeda. Regardless of their associations, the United States ultimately devoted the same resources to picking off locally affiliated militants as it did to the campaign against the group responsible for 9/11.

After nearly a decade of war, thousands of operations, and thousands of deaths, some within the special operations community began to question the quality of the Americans' targets in Afghanistan. "By 2010, guys were going after street thugs," a former SEAL Team 6 officer told the *New York Times* in June 2015. "The most highly trained force in the world, chasing after street thugs."[6] Concerns that the United States was devoting tremendous resources to kill a never-ending stream of nobodies did little to halt the momentum.

The secret Haymaker documents, which include a slide on "Manhunting Basics," reflect the combination of U.S. military person-

Cyber and military intelligence specialists monitor U.S. Army networks in the Cyber Mission Unit's Cyber Operations Center at Fort Gordon, Georgia.

nel and spies who have hunted targets along Afghanistan's border with Pakistan for years. According to one of the slides, the Haymaker "functional teams" included the CIA, the Defense Intelligence Agency, the NSA, and the National Geospatial-Intelligence Agency. The air force's uniquely designed 11th Intelligence Squadron also played a role. The Florida-based squadron was reactivated in August 2006 for the express purpose of supporting "find, fix, finish" operations to capture or kill targets through analysis of aerial intelligence.[7]

The drone operations that supported campaigns like Haymaker also included personnel stationed at Camp Alpha, a secure facility at Bagram populated by teams from the military's Joint Special Operations Command, as well as contractors manning stations at U.S. bases like Fort Gordon, a lesser-known though crucial node in the war on terror that supports tactical NSA operations abroad from Augusta, Georgia. The hunting and killing operations relied on advanced technology to zero in on targets, including the cell phone

SECRET//NOFORN

OPN HAYMAKER
Functional Teams

Current Ops (24hr cycle)
Targeting Officer
NGA analyst
SIGINT analysts
ITCs
FMV screeners

HAT
D7/D13
11th IS
CIA

Future Ops (48hr cycle)
SIGINT analysts
Intel. analysts
FMV screeners

NSA-Washington
NSA-Georgia
CNOS

DIA Sr. analyst
NGA Targeteer

Development Ops (+48hr cycle)

36 total personnel attached to TF EAST

A slide detailing Operation Haymaker's "functional teams," which included personnel from a range of U.S. military units and intelligence agencies.

geolocation system known as Gilgamesh, which employs a simulated cell phone tower to identify and locate targeted SIM cards.

While signals intelligence and electronic surveillance clearly contributed to Haymaker's kinetic operations, such as the drone strike that targeted Qari Munib, there was evidently more to the missions than advanced technology. Unlike some other arenas in which the war on terror has touched down – Yemen and Somalia, for example – the Haymaker documents point to the robust presence of U.S. intelligence agencies and human sources on the ground in northeastern Afghanistan. In Nuristan's Waygal district, the Defense Intelligence Agency and the CIA had "myriad complementary sources," notes a slide laying out "targeting criteria" in the district, "some of whom may be able to trigger our operations." A third of the "active targets" in Waygal had "good selectors" – phones the Americans could target in the run-up to a raid or airstrike – and the NSA had "taken on [signals intelligence] development in Waygal, greatly enhancing our understanding of the [signals intelligence] environment."

SECRET//NOFORN

Manhunting Basics

- **HUMINT** These aren't the terrorist you're looking for
- **IMINT** FMV is your friend ... and enemy
- **SIGINT** May I ask who's calling, please?
- **ABI** More layers are preferred
- **SNA** More connections aren't always best
- **IPB** Yep, IPB
- **F3EA** Can we stop adding letters?

- **Kinetic v. Direction Action**

The slide titled "Manhunting Basics" takes a lighthearted approach to the core mission of the Haymaker campaign: finding and killing specific individuals.

According to the slide, Waygal, described as a "historic" al Qaeda sanctuary, included more than a half-dozen NAIs, "named areas of interest," the identification of which was attributed to similarly "historic" levels of human and signals intelligence cultivation, as well as surveillance provided by drones scanning the district. There were "over a dozen active targets" in three villages, and most of the targets were already on a targeting list, or "easily could be." "The targets there are not only senior-level Taliban facilitators and hosts, but Arabs themselves," the slide notes, underscoring the presence of suspected foreign fighters in the district. "Elimination of these targets will provide demonstrable measures of success."

The documents indicate that U.S. forces launched just one airstrike as part of the Haymaker campaign in the early months of 2012, killing two people. In May, however, the tempo of operations picked up dramatically, an increase that coincided with a strategic shift in Afghanistan emanating from the White House. As the mil-

OP WAYGAL – Operational Targeting Criteria

Waygal District:

- *Maturity of HUMINT network* – **HIGH**.
  - HI networks have developed over the past 12-18 months. HAT and OGA have myriad complementary sources, some of whom may be able to trigger our operations.
- *Quality of established NAIs* – **HIGH**.
  - We currently have high fidelity on 8x NAIs in Waygal. This speaks to the level of historic HI, SI, FMV development in the valley.
- *Depth of SIGINT start points* – MED.
  - A third of our active targets have good selectors. NSA-W has taken on SI development in Waygal, greatly enhancing our understanding of the SI environment.
- *Number of active targets* – **HIGH**.
  - Over a dozen active targets across 3 villages, most of whom are JTL'd or could easily be. Known AQ Arabs are on this deck.
- *H-value** – **HIGH**.
  - Waygal is an historic AQ sanctuary. The targets there are not only senior-level Taliban facilitators and hosts, but Arabs themselves.
  - Faruq al-Qahtani and Dost Mohammad and their entourages frequent Waygal.
  - Elimination of these targets will provide demonstrable measures of success.

HAYMAKER Value (H-Value) is the assessed return on investment of achieving success in a named operation

A slide reporting a robust presence of human intelligence sources in the district of Waygal, a "historic" al Qaeda sanctuary in northeastern Afghanistan.

itary's focus shifted to hunting down specific targets from 2011 to 2012, drone strikes in Afghanistan increased by 72 percent.[8]

Over the course of five months, stretching through the summer of 2012, Haymaker operations included twenty-seven raids and twenty-seven airstrikes. The raids resulted in the capture of sixty-one people, thirteen of them jackpots, the actual targets of the missions. A total of two people were reportedly killed in these ground operations. In the airstrikes, meanwhile, 155 people were killed and labeled as enemies killed in action, according to a table in the documents. Just nineteen were jackpots. The table does not say whether the jackpots are reflected in the EKIA total. It does, however, appear to present a success rate: the number of jackpots divided by the number of missions. In the case of raids, a figure of 48 percent is presented; for airstrikes it's 70 percent.

The scores of unnamed people killed in the hunt for jackpots and the intelligence opportunities lost by failing to capture targets live do not appear to factor into the calculation. The apparent success rate, in other words, depends solely on killing jackpots and

HAYMAKER Operations (01 May – 15 Sep 2012)					
Type	# Ops	EKIA	Detainees	JP	%
Enabled Ops	27	2	61	13	48%
Kinetic Strikes	27	155	N/A	19	70%
Total	54	157	61	32	

A comparison of raids (described as "enabled ops") and airstrikes (described as "kinetic strikes") reveals significant differences in the total number of prisoners taken versus individuals killed during an intensified period of Haymaker operations.

ignores the strategic – and human – consequences of killing large numbers of bystanders.

While the source conceded there could be scenarios in which women and children killed in an airstrike are labeled EKIA, in the case of the Haymaker strikes he believed it was "more likely" that the dead included "groups of men or teenage boys" killed because "the intel says the guy JSOC is going after may be in that group of men." In the event that a target is identified in such a group, he said, "they'll go through with the strike."

Also included is a chart revealing that airstrikes killed 219 people over a fourteen-month period in 2012 and 2013, resulting in at least thirty-five jackpots. The document includes thumbnail images of individuals, representing a range of groups, who were captured or killed during Haymaker: thirty men, twenty-four of them stamped EKIA, five detained, and one wounded in action. The deaths of just over half the individuals were noted in NATO's press releases or media reports.

The Haymaker files also point to the psychological impact of living under the constant threat of death from above – an effect human rights workers have documented among civilians living in areas populated by militants.[9] A quote attributed to a Taliban detainee identified as "Ahmad," also known as "Objective Brandywine," features prominently on three of the documents. "Hands down, the scariest/most intimidating message for the Taliban, at any level, from fighter to Taliban senior leadership, is anything to do with drones or aerial bombings," Ahmad purportedly said. "The Taliban has no way to defend against them and they are certain to end in absolute destruction of whatever their target is."

Still, the documents' assessment of Haymaker's effectiveness is frank. A slide detailing the campaign's "effects" from January 2012 through February 2013 includes an assessment of "Objectives & Measures of Effectiveness." The results were not good. Disruptions in al Qaeda's view of northeastern Afghanistan as a safe haven and the loss of "key" al Qaeda members and enablers in the region were deemed "marginal." Meanwhile a comparison of Haymaker 1.0 (August 2011) with Haymaker 2.0 (February 2013) notes that al Qaeda faced "little to no local opposition" and enjoyed "relatively free movement" to and from Pakistan. Kinetic strikes, the slide reports, "successfully killed one [al Qaeda] target per year," allowing the organization to "easily" reconstitute.

By 2013 Haymaker was amassing a significant body count but making little headway against al Qaeda forces in the region. According to the "Success Criteria," slide, "sporadic reporting of concern over [the] viability" of northeastern Afghanistan as a safe haven had been "overshadowed" by the group's senior leadership discussing the establishment of a "post-2014 sanctuary." Individuals continued to return to Pakistan to support operations in and outside of Afghanistan, the slide asserts. While "nascent developments in some valleys" indicated that locals were "tiring" of al Qaeda's efforts to "root out spies as a perceived method to stopping strikes," the strikes and raids themselves had "succeeded in killing/capturing few [al Qaeda] targets." As slides detailing its effectiveness note, Haymaker's impact on al Qaeda and Taliban enablers in Kunar and Nuristan was "considered temporary without a long-term, persistent campaign."

Senior Nuristan Taliban Detainee
Ahmad (OBJ BRANDYWINE)
"Hands down, the scariest/most intimidating message for the Taliban, at any level, from fighter to Taliban senior leadership, is anything to do with drones or aerial bombings. The Taliban has no way to defend against them and they are certain to end in absolute destruction of whatever their target is."

A quote attributed to an alleged Taliban detainee describes the psychological impact of living under the threat of U.S. airstrikes.

On February 18, 2013, while Haymaker was still under way, Afghan President Hamid Karzai issued a decree: Afghan military forces were barred from calling in U.S. airstrikes for support on missions.[10] The order followed an operation in Kunar in which NATO and Afghan forces were blamed for the deaths of ten civilians: one man, four women, and five children. In 2012 the UN mission in Afghanistan had documented a number of other incidents involving civilian deaths

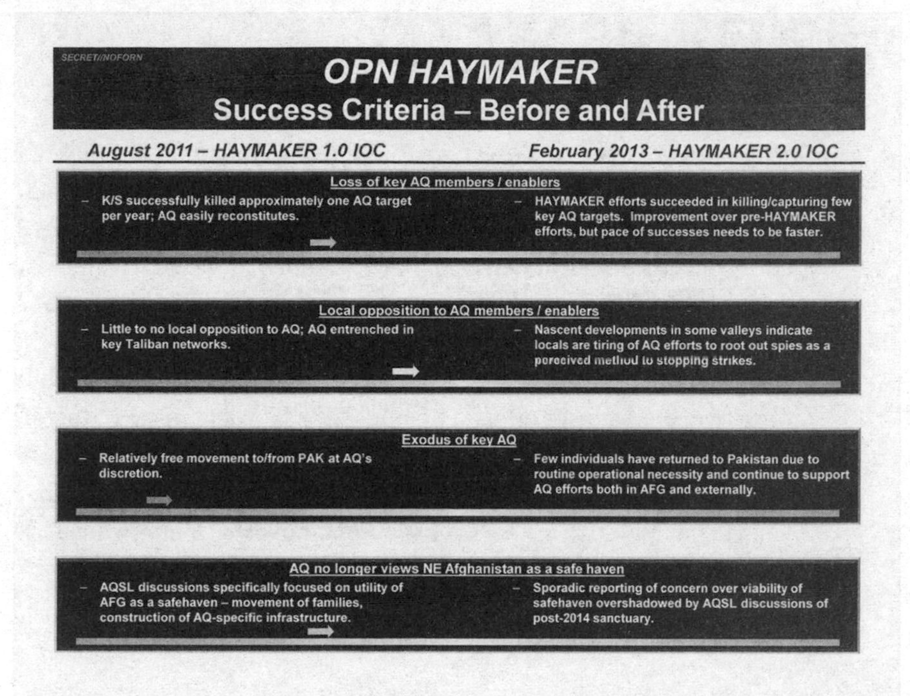

A slide reveals the Haymaker campaign's limited strategic achievements.

resulting from U.S. operations, including a raid that left seven civilians dead, an "aerial attack" that killed seven children and one adult, and a drone strike targeting "two insurgents" that killed a teenage girl.

The most recent date included in the Haymaker materials is February 28, 2013. Whether that date marked the end of the campaign is unclear. What does seem clear, however, is that Haymaker coincided with an increase in drone strikes and civilian casualties across Afghanistan. By the end of 2013 the UN reported that the number of civilian casualties from drone strikes in Afghanistan had tripled from 2012, with "almost one-third of the civilian deaths from aerial operations" reported in Kunar, the heart of the Haymaker campaign. Records of condolence payments disbursed by the U.S. military, obtained by *The Intercept*, show more than $118,000 distributed in forty-five disbursements to Kunar in fiscal years 2011 through 2013.[11] In addition to numerous injuries, the payments also cover compensation for the deaths of twenty-seven people, including at least four children. The records do not indicate whether the incidents were

AFGHAN OBJECTIVES

Documents obtained by *The Intercept* detailing Operation Haymaker, a U.S. military campaign carried out in the provinces of Kunar and Nuristan in northeastern Afghanistan, stretching from late 2011 into early 2013, identify dozens of individuals detained, wounded, or killed in American operations. This table contains the names, alleged militant affiliations, and ranks of 31 of those individuals, whose deaths were confirmed in the Haymaker materials, open source reports, or both. Though it is not comprehensive, the data set offers a unique glimpse at the range of targets elite U.S. forces have hunted along the Afghanistan-Pakistan border in recent years, from al Qaeda's second in command in Afghanistan, Sakhr al-Taifi, to fighters such as the Salafist commander Haji Matin, whose militant objectives were more closely tied to the presence of U.S. forces in Kunar province.

NAME	OBJECTIVE	GROUP	Rank_MW	EKIA
JAMIL UR-RAHMAN	LEAD HILL	TB/AQ FAC	Commander	EKIA
SAKHR AL-TAIFI	VILONIA	AQ	Deputy emir	EKIA
NABI RAHMAN	SKANG	AQ/TB	Commander & facilitator	EKIA
MUSA	POSTAL SERVICE	TB/AQ FAC	Commander & facilitator	EKIA
HANZALLAH	COLLISON	AQ	Commander	EKIA
YUSUF	MIRAMAX	AQ/HUJI	Commander	EKIA
MUFTI ASAD	TOUCHSTONE	AQ/HUJI	Commander	EKIA
ABU WALID	CHATHAM BARS	AQ	Commander	EKIA
FATEH GUL	VULCAN	AQ	Facilitator	EKIA
TURABI	GREEN RUN	TB/AQ FAC	Facilitator	EKIA
ATTIQULLAH	HUDSON	TB/AQ FAC	Commander	EKIA
MATAQI	POST MORTEM	TB/AQ FAC	Facilitator	EKIA
DADULLAH	RAYBURN	TTP	Commander	EKIA
ABU SAIF	LIONSGATE	AQ/HUJI	Commander	EKIA
MULLAH JALAL	ORONO	TB/AQ FAC		EKIA
ASADULLAH	RIBEYE	TB/AQ	Sub-Commander & facilitator	EKIA
ABDUL RAUF	WAKIKI	AQ/TTP	Sub-Commander & facilitator	EKIA
ASAD	KEITHSTONE	TTP/TB	Sub-Commander	EKIA
NUMAN	POSTMASTER	TB/AQ FAC	Commander	EKIA
YAHYA	JUNIOR BOYS	TB/AQ FAC	Facilitator	EKIA
MOHAMMAD RASUL	LETHAL ASPEN	TTP/TB	Commander	EKIA
SAIFULLAH	CANNON LAKE	TTP	Commander	EKIA
GHAFAR AHMAD	IOWA	TB	Commander	EKIA
HAMIDI	WIND RIVER	TB/AQ FAC	Sub-Commander & facilitator	EKIA
AMMAR	ANARCHY	LeT	Commander	EKIA
ABDULLAH	HEARTBREAK RIDGE	AQ	Commander	EKIA
DOST MOHAMMAD	DAKOTA	TB	Shadow governor	EKIA
TURAB aka KHANJAR	KNIFE GAME	AQ/TB	Commander & facilitator	EKIA
QARI MUNIB	LETHAL BURWYN	TB	Sub-Commander	EKIA
HAJI MATI	RONAN	Salafist	Commander	EKIA
DOWRA	VIKING	Salafist	Commander	EKIA

TB — Taliban
AQ — al Qaeda
AQ FAC — al Qaeda facilitator
HUJI — Harkat-ul-Jihad-al-Islami
LeT — Lashkar-e-Taiba
TTP — Tehrik-e Taliban Pakistan
EKIA — enemy killed in action

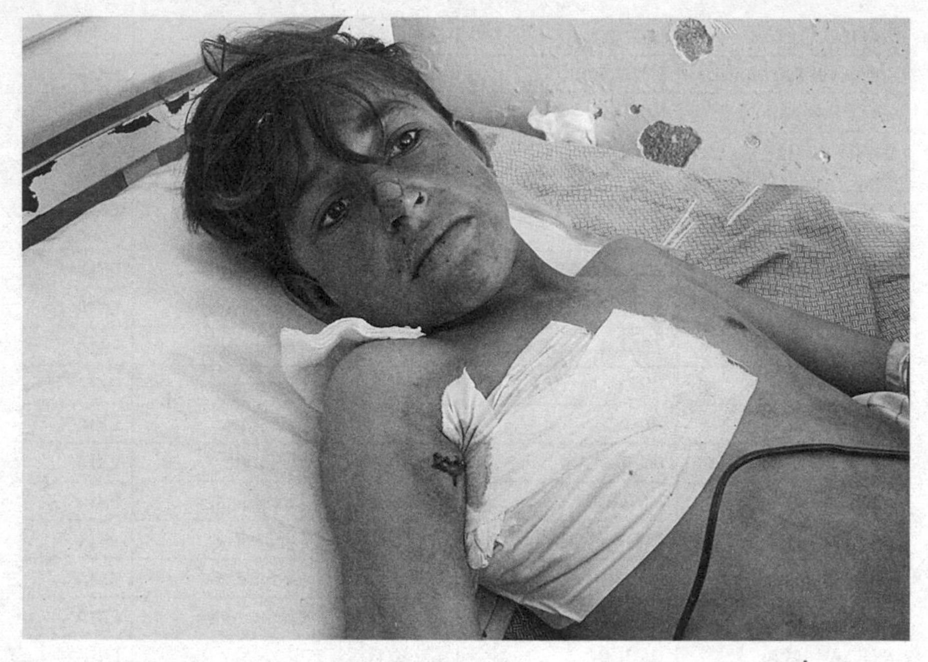

A wounded Afghan boy receives treatment at a hospital in Kunar Province on February 13, 2013, after a NATO airstrike killed ten civilians.

linked to the Haymaker campaign or were the result of mistaken ground raids or airstrikes.

Until recently the ongoing conflict in Afghanistan had largely receded from public conversations in the United States. But the American airstrike on a hospital run by the international organization Médecins Sans Frontières (Doctors Without Borders) offered a forceful reminder that the war, despite the Obama administration's declaration in 2014, is far from over. Unleashed in the early morning hours of October 3, 2015, in the province of Kunduz, the U.S. attack killed at least a dozen members of the humanitarian group's medical staff and ten patients, including three children. A nurse on the scene recalled seeing six victims in the intensive care unit ablaze in their beds. "There are no words for how terrible it was," the nurse said.[12] MSF denounced the strike as a war crime and demanded an independent investigation.

The Kunduz attack underscored an ugly reality: after nearly a decade and a half of war, more than 2,300 American lives lost, and an estimated 26,000 Afghan civilians killed, the nature of combat in Afghanistan is entering a new, potentially bloodier phase.[13] In August 2015 the United Nations reported that civilian casualties in Afghanistan were "projected to equal or exceed the record high numbers documented last year."[14] While most civilian casualties in the first half of 2015 were attributed to "anti-government" forces, twenty-seven deaths and twenty-two injuries were attributed to airstrikes "by international military forces," a 23 percent increase over the previous year, most of them, unlike the air raid in Kunduz, carried out by drones.

Despite the rise in civilian casualties and the well-documented failure of drone strikes to achieve the military's broader objectives, there is every indication that unmanned airstrikes will play an increasing role in U.S. military engagement in Afghanistan, as they have in war zones across the world. Less than two weeks after the UN issued its report, the journal *Foreign Policy* revealed that JSOC had drastically reduced the number of night raids it conducts in Afghanistan, while dramatically increasing its reliance on airstrikes, and was "on pace to double the rate at which it kills 'high-value individuals' using kinetic strikes, compared to how many it was killing that way five years ago."[15]

Afghanistan's northeastern border with Pakistan continues to be an active area of focus for the remaining U.S. special operations forces in the country. The Pech Valley, a hotspot during the Haymaker campaign, continues to host a constellation of armed groups. Al Qaeda, the organization used to justify both the invasion of Afghanistan and the Haymaker campaign, reportedly enjoys a more pronounced presence in the valley than ever. "The al Qaeda presence there now," according to a report by the U.S. Institute for Peace, "is larger than when U.S. counterterrorism forces arrived in 2002."[16]

With JSOC and the CIA running a new drone war in Iraq and Syria, much of Haymaker's strategic legacy lives on.[17] Such campaigns, the intelligence community source said, with their tenuous strategic impacts and significant death tolls, should serve as a reminder of the dangers fallible lethal systems pose. "This isn't to say that the drone program is a complete wash and it's never once succeeded in carrying out its stated purpose," he pointed out. "It certainly has." But even the operations that military commanders would point to as successes, he argued, can have unseen impacts, particularly in the remote communities where U.S. missiles so often rain down. “I would like to think that what we were doing was in some way trying to help Afghans,” the source explained, but the notion “that what we were part of was actually defending the homeland or in any way to the benefit of the American public” had evaporated long ago. “There’s no illusion of that that exists in Afghanistan. It hasn’t existed for many years.”

U.S. intelligence agencies hunt people primarily by locating their cell phone. Equipped with a simulated cell tower called Gilgamesh, a drone can force a target's phone to lock onto it and then use the phone's signals to triangulate that person's location.

Here is what a watchlist looks like.

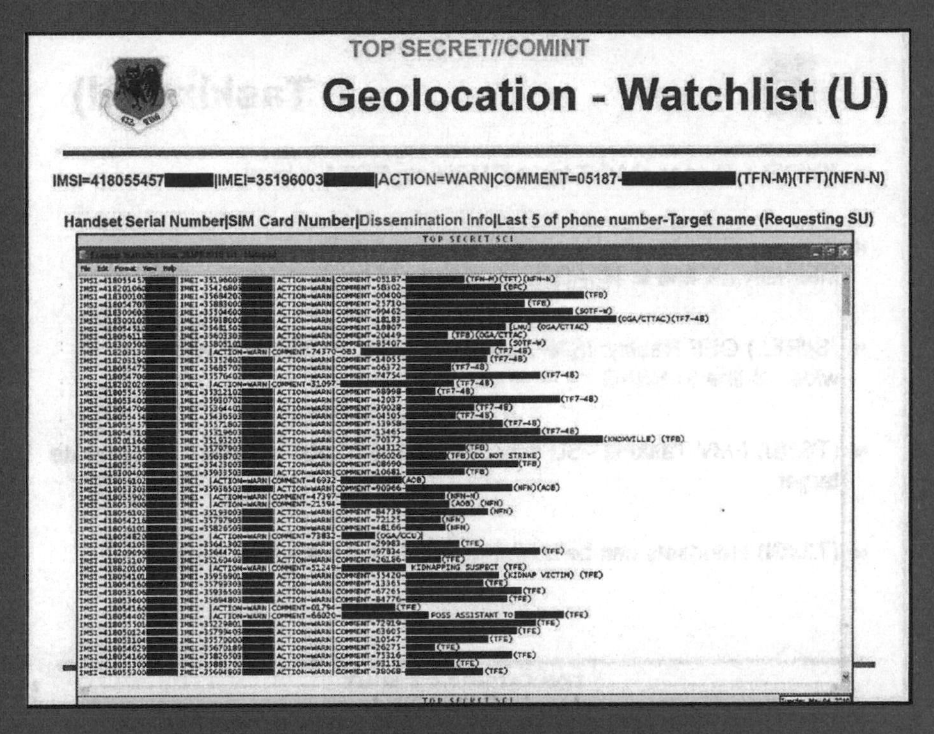

AFTERWORD: WAR WITHOUT END

GLENN GREENWALD

Barack Obama's 2008 presidential campaign is, for many, a distant memory. For that reason it is easy to forget that his vows to reverse the core strategies of the Bush-Cheney war on terror were central, not ancillary, to his electoral victory. Particularly when he was seeking the Democratic nomination, but even during the general election, candidate Obama railed with particular vehemence against the lawlessness that drove America's war policies since 9/11. "No more ignoring the law when it is inconvenient. That is not who we are," he thundered during a 2007 speech. "We will again set an example for the world" that "justice is not arbitrary."

The animating principle of Obama's promised new war-fighting approach would be restoration of international and constitutional law. A November 2010 profile of Attorney General Eric Holder in *GQ* emphasized its centrality not only to his 2008 campaign but also to the overall Obama brand of "change": "No other issue defined Barack Obama like his promise to restore America's commitment to international law. Other items may have topped his domestic agenda, but as a symbol of what Obama's candidacy *meant*, of what his election signified to the world nothing conveyed his message of 'change' like the pledge to repair American justice." As PolitiFact put it in 2013, "In the years leading up to the 2008 campaign and in the early stages of the campaign itself, Obama regularly emphasized the importance of civil liberties and the sanctity of the Constitution."

As is typical for Obama, much of his rhetoric in these areas was long on pretty language and short on specifics and substance. But there was one specific, concrete position on which he was particularly insistent: that it is inherently unjust for the U.S. government to treat individuals as terrorists – and, worse, to punish them as terrorists – without first providing them due process in the form of judicial review.

In September 2006 the U.S. Senate voted to pass the Military Commissions Act, which legalizes the military commissions created by the Bush administration at Guantánamo. Those commissions were empowered to punish detainees accused of terrorism. What made them so controversial, so radical from the perspective of American justice, was the full-scale denial of detainees' access to actual American courts. As a result those accused of terrorism offenses could be subjected to the harshest punishments, including long-term prison sentences, without any opportunity whatsoever to have a court review their claims of innocence.

Prior to the Senate vote, an amendment was proposed that would have vested accused terrorists with the right to have their case reviewed by a U.S. federal court. Senator Obama went to the Senate floor and spoke in passionate and deeply personal terms in support of this amendment, insisting that it was vital to force the government to prove a person's guilt before treating him or her as a terrorist:

> The bottom line is this: Current procedures . . . are such that a perfectly innocent individual could be held and could not rebut the Government's case and has no way of proving his innocence.
>
> I would like somebody in this Chamber, somebody in this Government, to tell me why this is necessary. I do not want to hear that this is a new world and we face a new kind of enemy. I know that. . . . But as a parent, I can also imagine the terror I would feel if one of my family members were rounded up in the middle of the night and sent to Guantanamo without even getting one chance to ask why they were being held and being able to prove their innocence.
>
> This is not just an entirely fictional scenario, by the way. We have already had reports by the CIA and various generals over the last few years saying that many of the detainees at Guantanamo should not have been there. . . .
>
> This is an extraordinarily difficult war we are prosecuting against

> terrorists. There are going to be situations in which we cast too wide a net and capture the wrong person.
>
> But what is avoidable is refusing to ever allow our legal system to correct these mistakes. By giving suspects a chance—even one chance—to challenge the terms of their detention in court, to have a judge confirm that the Government has detained the right person for the right suspicions, we could solve this problem without harming our efforts in the war on terror one bit. . . .
>
> For people who are guilty, we have the procedures in place to lock them up. That is who we are as a people. We do things right, and we do things fair. . . .
>
> Mr. President, this should not be a difficult vote. I hope we pass this amendment because I think it is the only way to make sure this underlying bill preserves all the great traditions of our legal system and our way of life.

Senator Obama repeatedly applied the same principles to Bush's surveillance policies. In 2006 he announced his opposition to Bush's appointment of the former NSA chief Gen. Michael Hayden to lead the CIA. Obama's opposition was based on Hayden's 2002 implementation of Bush's warrantless surveillance program, which Obama viewed as both illegal and unethical. Obama was particularly offended, as he put it, that "the National Security Agency has been spying on Americans without judicial approval." Justifying his vote against Hayden's confirmation, he inveighed, "What protects us, and what distinguishes us, are the procedures we put in place to protect that balance, namely judicial warrants and congressional review." We must, said Obama, "allow members of the other two coequal branches of government – Congress and the Judiciary – to have the ability to monitor and oversee such a program."

That was Obama explaining why mere *eavesdropping on* and *imprisonment of* suspected terrorists without due process were inherently abusive as well as violations of "all the great traditions of our legal system and our way of life." Back in 2006, as a senator, he understood that judicial procedures were necessary before treating someone as a terrorist because otherwise "we cast too wide a net" and end up accusing "the wrong person."

Yet as president, Obama not only ignored those lofty statements but trampled all over them. The centerpiece of his drone assassination program is that he, and he alone, has the power to target people, including American citizens, anywhere they are found in the world and order them executed on his unilateral command, based on his determination that the person to be killed is a terrorist. Somehow it was hideously wrong for George W. Bush to *eavesdrop on* and *imprison* suspected terrorists without judicial approval, yet it was perfectly permissible for Obama to *assassinate* them without due process of any kind.

It is hard to overstate the conflict between Obama's statements before he became president and his presidential actions. Bush's CIA and NSA chief Hayden said about Obama's drone assassination of Anwar al Awlaki, "We needed a court order to eavesdrop on him but we didn't need a court order to kill him. Isn't that something?"

It is indeed something. What it is in particular is a continuation, and in many cases an aggressive expansion, of the core principles of the Bush-Cheney mentality that Obama repeatedly vowed to overturn.

That Obama would embrace rather than repudiate these Bush-Cheney "war on terror" principles became evident almost immediately after he was inaugurated. Within the first several weeks of his presidency, his top legal officials explicitly advocated several of the most extremist and controversial theories of power that defined the Bush administration's approach to terrorism.

In February 2009 Obama's lawyers argued that detainees held in Bagram, Afghanistan, have no legal right to challenge their detention, thus, as characterized by the *New York Times*, "embracing a key argument of former President Bush's legal team." When a Bush-appointed federal judge rejected that claim, and held that—as Obama had once argued—detainees captured away from a battlefield deserve access to a court, the Obama Justice Department appealed the ruling, thus demonstrating that the administration "was not backing down in its effort to maintain the power to imprison terrorism suspects for extended periods without judicial oversight." This time the Obama administration won the right to imprison people without charges under the "war on terror" banner.

By 2010 that theory was fully extended to Guantánamo. In May 2009 President Obama delivered a speech at the National Archives, in front of the U.S. Constitution, proposing a system of preventative "prolonged detention" without trial inside the United States. In an article titled "President's Detention Plan Tests American Legal Tradition," the *New York Times* called Obama's plan "a departure from the way this country sees itself, as a place where people in the grip of the government either face criminal charges or walk free." In January 2010 the Obama administration announced it would continue to imprison several dozen Guantánamo detainees without charge or trial, including even a military commission, on the ground that these prisoners were "too difficult to prosecute but too dangerous to release." In other words, long-term detention without due process would now be acceptable for accused terrorists under President Obama.

Also in the first two months after Obama's inauguration, his administration invoked one of the most controversial Bush-Cheney legal weapons: the state secrets privilege. Conceived as a tool to allow the government, in the rarest of cases, to prevent the use of particularly sensitive documents in litigation, the Bush administration had embraced a version of the privilege that was warped beyond recognition: not only could specific documents be suppressed if they were sufficiently secret, but entire lawsuits alleging illegal government conduct could be dismissed at the start if the "subject matter" was a state secret.

When Bush administration lawyers used this doctrine to shield its torture, rendition, and surveillance programs from judicial review, Democrats, including Obama, vehemently denounced it. Indeed Obama's campaign website, Plan to Change Washington, contended in the section titled "The Problem" that the Bush administration "has invoked a legal tool known as the 'state secrets' privilege more than any other previous administration to get cases thrown out of civil court."

Yet in February 2009 the Obama Justice Department itself used this privilege to demand that a federal court dismiss a lawsuit brought by a victim of the U.S. extraordinary rendition program. The *New York Times* reported in an article headlined "On State Secrets, Obama Is Sounding Like Bush" that a Justice Department lawyer "seemed to surprise a panel of federal appeals judges on Monday by pressing ahead with an argument for preserving state secrets originally developed by the Bush administration."

Two months later Obama lawyers were back in court raising the same argument, this time to demand dismissal of a lawsuit challenging the legality and constitutionality of the NSA's warrantless eavesdropping program. In 2010 the Obama Justice Department used the doctrine in the most extreme way possible: to demand dismissal of a lawsuit brought by Awlaki's father that requested that a federal judge enjoin the president from killing his son without a trial. Rather than contest the lawsuit on the merits – by, for instance, showing evidence of Awlaki's guilt – Obama lawyers insisted that courts have no role to play in reviewing the president's war on terror killings.

Early developments like these led the *New York Times* Pulitzer Prize–winning reporter Charlie Savage to observe on February 17, 2009 – less than a month after Obama's inauguration – that "the Obama administration is quietly signaling continued support for . . . major elements of its predecessor's approach to fighting al Qaeda," which was "prompting growing worry among civil liberties groups and a sense of vindication among supporters of Bush-era policies." The headline said it all: "Obama's War on Terror May Resemble Bush's in Some Areas."

So unthinkable was this development – that Obama, fresh off a campaign railing against the Bush-Cheney war on terror, would then adopt and extend its defining provisions – that many refused to believe it was happening despite the mounting proof. I was one such person: in February I mildly objected to Savage's comparison of Bush and Obama as "premature." But by July the evidence was indisputable, and I wrote that I had been wrong and that Savage's article was "more prescient than premature."

By that point the realization that Obama's "war on terror" theories would largely track Bush's grew rapidly. Even right-wing Republicans, who had anticipated attacking Obama for abandoning the Bush-Cheney approach, were instead acknowledging, and praising, the continuity. Gen. Hayden, whose confirmation as CIA chief Obama opposed, gushed with praise for Obama: "There's been a powerful continuity between the 43rd and the 44th president." James Jay Carafano, a homeland security expert at the conservative Heritage Foundation, told the *New York Times* reporter Peter Baker, "I don't think it's even fair to call it Bush Lite. It's Bush. It's really, really hard to find a difference that's meaningful and not atmospheric." In a 2011 interview with NBC News, Dick Cheney himself observed, "He obviously has been through the fires of becoming President and having to make decisions and live with the consequences. And it's different than being a candidate. When he was candidate he was all for closing Gitmo. He was very critical of what we'd done on the counterterrorism area to protect America from further attack and so forth. . . . I think he's – in terms of a lot of the terrorism policies – the early talk, for example, about prosecuting people in the CIA who've been carrying out our policies – all of that's fallen by the wayside. I think he's learned that what we did was far more appropriate than he ever gave us credit for while he was a candidate. So I think he's learned from experience."

Obama did not navigate this transformation alone. As is to be expected in the highly partisan and polarized political climate that prevails in the United States, large numbers of Democrats and progressives transformed with him from virulent critics of these policies to vocal supporters once they became Obama policies rather than Bush policies. A 2012 poll from the *Washington Post*/ABC News found that "the sharpest edges of President Obama's counterterrorism policy, including the use of drone aircraft to kill suspected terrorists abroad and keeping open the military prison at Guantánamo Bay, have broad public support, including from the left wing of the Democratic Party." Indeed "53 percent of self-identified liberal Democrats – and 67 percent of moderate or conservative Democrats – support keeping Guantánamo Bay open, even though it emerged as a symbol of the post–Sept. 11 national security policies of George W. Bush, which many liberals bitterly opposed." More amazing still, "fully 77 percent of liberal Democrats endorse the use of drones," and "support for drone strikes against suspected terrorists stays high, dropping only somewhat when respondents are asked specifically about targeting American citizens living overseas, as was the case with Anwar al Awlaki."

A former Bush Justice Department lawyer and current Harvard Law professor, Jack Goldsmith, wrote in the *New Republic*, "The new administration has copied most of the Bush program, has expanded some of it, and has narrowed only a bit." He highlighted the critical point: that Obama made those policies bipartisan and thus *strengthened* them beyond what Bush and Cheney could ever have achieved on their own. Goldsmith argued that Obama's decision "to continue core Bush terrorism policies is like Nixon going to China"; in other words, only a liberal, self-proclaimed civil libertarian president like

Obama had the ability to institutionalize these once-controversial and radical powers into mainstream, bipartisan powers.

Obama's aggressive, expansive use of drones over the course of seven years, in multiple, predominantly Muslim countries, embodies the worst of what made the Bush-Cheney "war on terror" approach so destructive. Obama arrogated unto himself the unilateral, unrestrained, unchecked power to decide – without a whiff of due process – who is a terrorist and who should die. The extension of this claimed authority to an American citizen was a serious escalation of what even Bush and Cheney dared to do. Indeed Bush officials remarked, almost certainly accurately, that they would never have gotten away with killing Awlaki the way Obama did.

There were, of course, a couple of "war on terror" changes as a result of Obama's election: the torture program (which had already been deactivated before the 2008 election) formally ended, and special forces and drones replaced the type of large-scale ground invasion that destroyed Iraq. But the defining essence of the Bush-Cheney template – that the United States is fighting an endless war against *terror suspects* who have no due process rights of any kind – is very much alive and, in many cases, stronger than ever.

The Assassination Complex provides the most in-depth look into how these powers have been used and abused. It powerfully validates Obama's prepresidency concern about unchecked action against terrorists: that "there are going to be situations in which we cast too wide a net and [target] the wrong person." Most of all the revelations in this book signify one of the most enduring and consequential aspects of the Obama legacy: the continuation of endless war, fueled by the very powers he was elected vowing to end.

177←

FIND, FIX, FINISH

→13

In the end, Intelligence, Surveillance, and Reconnaissance (ISR) is about continuing a cycle: Find the person. Fix the person. Finish the person. Two other steps in the process are Exploit and Analyze.

Colloquially referred to as "F3EA," the cycle feeds back into itself. The whole process amounts to human hunting. As soon as a target is finished, the hunt for a new target begins.

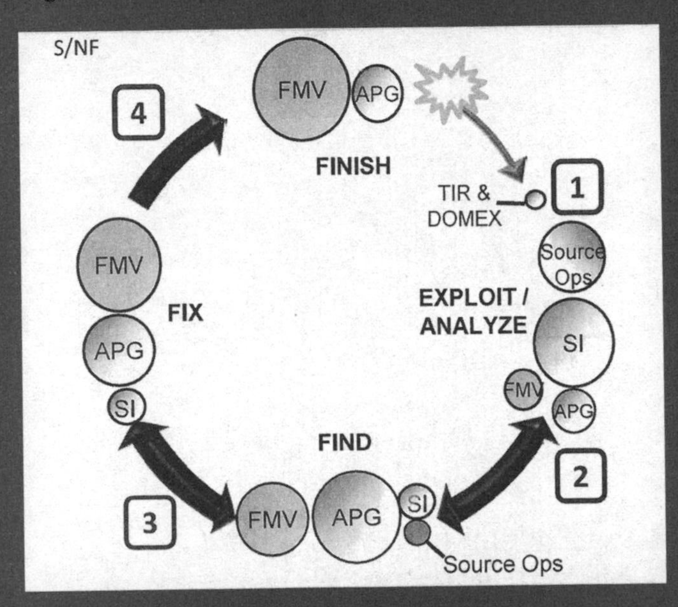

ACKNOWLEDGMENTS

This volume is the product of a collective journalistic enterprise, conducted over many months and involving *The Intercept*'s entire staff. While everyone else kept the website pulsing, the core "Drone Papers" team of reporters, whose work is featured in this volume, toiled under strict security constraints. Betsy Reed and Roger Hodge edited the original series, with contributions from Peter Maass, Ryan Tate, Andrea Jones, Rubina Madan Fillion, Miriam Pensack, and Charlotte Greensit. Thanks to Laura Poitras, Henrik Moltke, Andrew Fishman, and Matthew Cole for additional reporting. Crucial research was provided by Lynn Dombek, Sheelagh McNeill, Josh Begley, Margot Williams, Alleen Brown, John Thomason, and Spencer Woodman. With coding by engineers Tom Conroy and Raby Yuson, and photo editing by Connie Yu, Stéphane Elbaz and Philipp Hubert transformed complex technical content into captivating digital designs, and Philipp then expertly ushered that iconic imagery into book form. Lynn Oberlander provided unerring legal counsel, asking tough questions that made the journalism stronger. We are also grateful to Anthony Arnove, the agent who saw the potential for this book and did not rest until it was fulfilled; to Ben Wyskida and BerlinRosen for ensuring the series received maximum exposure; to Michael Bloom and the whole First Look Media team, from product to technology, security, communications, and accounting, for their unwavering support; and to Pierre Omidyar, for allowing us to pursue our journalism wherever it may take us. That freedom is an extraordinary gift for which we are deeply grateful.

Finally, we extend our deepest thanks to the brave whistleblower who at great personal risk provided the classified documents on the U.S. assassination program.

THE DOCUMENTS

"The Drone Papers" and all supporting documents are available at https://theintercept.com/drone-papers/.

ISR Task Force Requirements and Analysis Division, "ISR Support to Small Footprint CT Operations – Somalia/Yemen," February 2013, published by *The Intercept*, October 15, 2015, https://theintercept.com/document/2015/10/15/small-footprint-operations-2-13/.

ISR Task Force Requirements and Analysis Division, "ISR Support to Small Footprint CT Operations – Somalia/Yemen: Executive Summary," May 2013, published by *The Intercept*, October 15, 2015, https://theintercept.com/document/2015/10/15/small-footprint-operations-5-13/.

"Operation Haymaker," published by *The Intercept*, October 15, 2015, https://theintercept.com/document/2015/10/15/operation-haymaker/.

"Geolocation Watchlist," published by *The Intercept*, October 15, 2015, https://theintercept.com/document/2015/10/15/geolocation-watchlist/.

Other documents published by *The Intercept* and referenced in this volume:

March 2013, "Watchlisting Guidance," published by *The Intercept*, July 23, 2014, https://theintercept.com/document/2014/07/23/march-2013-watchlisting-guidance/.

"Directorate of Terrorist Identities (DTI): Strategic Accomplishments 2013," published by *The Intercept*, August 5, 2014, https://theintercept.com/document/2014/08/05/directorate-terrorist-identities-dti-strategic-accomplishments-2013/.

"The Secret Surveillance Catalogue," published by *The Intercept*, December 17, 2015, https://theintercept.com/surveillance-catalogue/.

Every document ever published by *The Intercept* can be found at https://theintercept.com/documents/.

NOTES

THE DRONE LEGACY

1. Elizabeth B. Bazan, "Assassination Ban and E.O. 12333: A Brief Summary," CRS Report for Congress, January 4, 2002, http://fas.org/irp/crs/RS21037.pdf.
2. William Saletan, "Editors for Predators: To Justify Drone Strikes, the Obama Administration Is Twisting Language and the Law," *Slate*, February 8, 2013, http://www.slate.com/articles/news_and_politics/frame_game/2013/02/drones_law_and_imminent_attacks_how_the_u_s_redefines_legal_terms_to_justify.html.
3. "CIA 'Killed al-Qaeda Suspects' in Yemen," *BBC News World Edition*, November 5, 2002, http://news.bbc.co.uk/2/hi/2402479.stm.
4. White House, "U.S. Policy Standards and Procedures for the Use of Force in Counterterrorism Operations Outside the United States and Areas of Active Hostilities," May 23, 2013. https://www.whitehouse.gov/sites/default/files/uploads/2013.05.23_fact_sheet_on_ppg.pdf.
5. Anthony D. Romero, "ACLU Comment on President's National Security Speech," ACLU, May 23, 2013, https://www.aclu.org/news/aclu-comment-presidents-national-security-speech.
6. ISR Task Force Requirements and Analysis Division, "ISR Support to Small Footprint CT Operations – Somalia/Yemen" (February 2013), *The Intercept*, October 15, 2015, https://theintercept.com/document/2015/10/15/small-footprint-operations-2-13/; ISR Task Force Requirements and Analysis Division, "ISR Support to Small Footprint CT Operations – Somalia/Yemen: Executive Summary" (May 2013), *The Intercept*, October 15, 2015, https://theintercept.com/document/2015/10/15/small-footprint-operations-5-13/.
7. "Operation Haymaker," *The Intercept*, October 15, 2015, https://theintercept.com/document/2015/10/15/operation-haymaker/. See also Jo Becker and Scott Shane, "Secret 'Kill List' Tests Obama's Principles," *New York Times*, May 29, 2012, http://www.nytimes.com/2012/05/29/world/obamas-leadership-in-war-on-al-qaeda.html?_r=0.
8. "Geolocation Watchlist," *The Intercept*, October 15, 2015, https://theintercept.com/document/2015/10/15/geolocation-watchlist/.
9. House Permanent Select Committee on Intelligence, "Performance

Audit of Department of Defense Intelligence, Surveillance, and Reconnaissance," April 2012, http://intelligence.house.gov/sites/intelligence.house.gov/files/documents/ISRPerformanceAudit%20Final.pdf.

10. Greg Miller, "U.S. Launches Secret Drone Campaign to Hunt Islamic State Leaders in Syria," *Washington Post*, September 1, 2015, https://www.washingtonpost.com/world/national-security/us-launches-secret-drone-campaign-to-hunt-islamic-state-leaders-in-syria/2015/09/01/723b3e04-5033-11e5-933e-7d06c647a395_story.html.
11. "The Drone Papers," *The Intercept*, October 15, 2015, https://theintercept.com/drone-papers/.
12. Scott Shane, "Drone Strikes Reveal Uncomfortable Truth: U.S. Is Often Unsure about Who Will Die," *New York Times*, April 23, 2015, http://www.nytimes.com/2015/04/24/world/asia/drone-strikes-reveal-uncomfortable-truth-us-is-often-unsure-about-who-will-die.html?mtrref=undefined&gwh=56B01FA77C9445466FEE3092EEF7E6A0&gwt=pay&assetType=nyt_now; Adam Entous, Damian Paletta, and Felicia Schwartz, "American, Italian Hostages Killed in CIA Drone Strike in January," *Wall Street Journal*, April 23, 2015, http://www.wsj.com/articles/american-italian-hostages-killed-in-cia-drone-strike-in-january-1429795801.

DEATH AND THE WATCHLIST

1. "March 2013 Watchlisting Guidance," *The Intercept*, July 23, 2014, https://theintercept.com/document/2014/07/23/march-2013-watchlisting-guidance/.
2. Dec. of Eric H. Holder, JR. Exhibit 1, Def Mem Support of Motion to Dismiss, Mohamed v. Holder, No. 1:11-CV-0050 (E.D. Va 5/28/2014), https://www.documentcloud.org/documents/1232048-holder-on-watchlist.html.
3. Bob Orr, "Inside a Secret U.S. Terrorist Screening Center," *CBS News*, October 1, 2012, http://www.cbsnews.com/news/inside-a-secret-us-terrorist-screening-center/.
4. Rachel L. Swarns, "Senator? Terrorist? A Watch List Stops Kennedy at Airport," *New York Times*, August 20, 2004, http://www.nytimes.com/2004/08/20/us/senator-terrorist-a-watch-list-stops-kennedy-at-airport.html.
5. Steve Kroft, "Unlikely Terrorists on No Fly List," *CBS News*, Octo-

ber 5, 2006, http://www.cbsnews.com/news/unlikely-terrorists-on-no-fly-list/.

6. Lizette Alvarez, "Mikey Hicks, 8, Can't Get Off U.S. Terror Watch List," *New York Times*, January 13, 2010, http://www.nytimes.com/2010/01/14/nyregion/14watchlist.html.
7. Federal Bureau of Investigation, "Office of the Inspector General Semiannual Report to Congress: April 1, 2007–September 30, 2007," https://oig.justice.gov/semiannual/0711/fbi.htm.
8. White House, Office of the Press Secretary, "Remarks by the President on Strengthening Intelligence and Aviation Security," January 7, 2010, https://www.whitehouse.gov/the-press-office/remarks-president-strengthening-intelligence-and-aviation-security.
9. U.S. Government Accountability Office, Report to Congressional Requesters, "Terrorist Watchlist: Routinely Assessing Impacts of Agency Actions since the December 25, 2009, Attempted Attack Could Help Inform Future Efforts," May 2012, http://www.gao.gov/assets/600/591312.pdf.
10. Defendants' Objections and Responses to Plaintiff's First Set of Interrogatories, Exhibit B, Mohamed v. Holder, No. 1:11-CV-0050, E.D. Va., April 7, 2014, https://www.documentcloud.org/documents/1232171-91-3.html.
11. Opinion and Ruling, Latif v. Holder, No. 3:10-CV-00750-BR, D. Ore., June 24, 2014. https://www.aclu.org/sites/default/files/assets/no_fly_list_ruling__-_latif_v._holder_-_6-24-14.pdf.
12. "Directorate of Terrorist Identities (DTI): Strategic Accomplishments 2013," *The Intercept*, August 5, 2014, https://theintercept.com/document/2014/08/05/directorate-terrorist-identities-dti-strategic-accomplishments-2013/.
13. Matthew Barakat, "US Terrorist Database Growing at Rapid Rate," *Yahoo News*, July 18, 2014, https://news.yahoo.com/us-terrorist-database-growing-rapid-rate-223303875.html?soc_src=copy.

FIND, FIX, FINISH

1. Scott Shane, "C.I.A. to Expand Use of Drones in Pakistan," *New York Times*, December 3, 2009, http://www.nytimes.com/2009/12/04/world/asia/04drones.html?mtrref=undefined&gwh=1FA2AA70F82914DE99DD705B558572FE&gwt=pay.
2. Mark Mazzetti and Dexter Filkins, "U.S. Military Seeks to Expand

Raids in Pakistan," *New York Times*, December 20, 2010, http://www.nytimes.com/2010/12/21/world/asia/21intel.html?mtrref=undefined&gwh=67B7EC85E9E45D1065C06C6680153771&gwt=pay; Chris Woods, "CIA's Pakistan Drone Strikes Carried Out by Regular US Air Force Personnel," *Guardian*, April 14, 2014, http://www.theguardian.com/world/2014/apr/14/cia-drones-pakistan-us-air-force-documentary.

3. Mark Mazzetti, "U.S. Is Said to Expand Secret Actions in Mideast," *New York Times*, May 24, 2010, http://www.nytimes.com/2010/05/25/world/25military.html?mtrref=undefined&gwh=4879E16FFEA45B298AA32E3C6AA7CA98&gwt=pay.
4. Peter Bergen and Jennifer Rowland, "Obama Ramps Up Covert War in Yemen," *CNN*, June 12, 2012, http://www.cnn.com/2012/06/11/opinion/bergen-yemen-drone-war/.
5. Jeremy Scahill, "Washington's War in Yemen Backfires," *The Nation*, February 15, 2012; "Al-Majalah Freedom of Information Act Request," Center for Constitutional Rights, April 17, 2012.
6. "CIA 'Killed al-Qaeda Suspects' in Yemen," *BBC News World Edition*, November 5, 2002, http://news.bbc.co.uk/2/hi/2402479.stm.
7. Karen DeYoung, "Special Forces Raid in Somalia Killed Terrorists With Al-Qaeda Links, U.S. Says," *Washington Post*, September 15, 2009, http://www.washingtonpost.com/wp-dyn/content/article/2009/09/14/AR2009091403522.html; Ken Dilanian, "U.S. Holds Somali Extremist for Two Months on Ship," *Seattle Times*, July 5, 2011, http://www.seattletimes.com/nation-world/us-holds-somali-extremist-for-two-months-on-ship/.
8. Jo Becker and Scott Shane, "Secret 'Kill List' Tests Obama's Principles," *New York Times*, May 29, 2012, http://www.nytimes.com/2012/05/29/world/obamas-leadership-in-war-on-al-qaeda.html?mtrref=undefined&gwh=DECAEF276E039B9D57EF3E8DBC75FF8F&gwt=pay.
9. Siobhan Gorman, Adam Entous, and Julian E. Barnes, "U.S. to Shift Drone Command," *Wall Street Journal*, March 20, 2013, http://www.wsj.com/articles/SB10001424127887324103504578372703357207828.
10. House Permanent Select Committee on Intelligence, "Performance Audit of Department of Defense Intelligence, Surveillance, and Reconnaissance," April 2012, http://intelligence.house.gov/sites/intelligence.house.gov/files/documents/ISRPerformanceAudit%20Final.pdf.

11. Donna Miles, "Gates Forms Task Force to Promote Intelligence, Surveillance for Warfighters," U.S. Department of Defense, April 21, 2008, http://archive.defense.gov/news/newsarticle.aspx?id=49639.
12. White House, Office of the Press Secretary, "Remarks by the President at the National Defense University," May 23, 2013, https://www.whitehouse.gov/the-press-office/2013/05/23/remarks-president-national-defense-university.
13. Karen DeYoung, "CIA Veteran John Brennan Has Transformed U.S. Counterterrorism Policy," *New York Times*, October 24, 2012, https://www.washingtonpost.com/world/national-security/cia-veteran-john-brennan-has-transformed-us-counterterrorism-policy/2012/10/24/318b8eec-1c7c-11e2-ad90-ba5920e56eb3_story.html.
14. Jeremy Scahill, "Inside America's Dirty Wars," *The Nation*, April 24, 2013, http://www.thenation.com/article/inside-americas-dirty-wars/.
15. Michael S. Repass, "Combating Terrorism with Preparation of the Battlespace," USAWC Strategy Research Project, April 1, 2003, http://fas.org/man/eprint/respass.pdf.
16. Eric Schmitt, "Pentagon Sends Its Spies to Join Fight on Terror," *New York Times*, January 24, 2005, http://www.nytimes.com/2005/01/24/politics/pentagon-sends-its-spies-to-join-fight-on-terror.html?mtrref=undefined&gwh=9C83AE763947310CED2C2AE0F07FB523&gwt=pay; "SEAL Team 6: A Secret History of Quiet Killings and Blurred Lines," *New York Times*, June 6, 2015, http://www.nytimes.com/2015/06/07/world/asia/the-secret-history-of-seal-team-6.html?_r=0&mtrref=undefined&gwh=1BFC4755DB8744A1FD49682A1F424253&gwt=pay.
17. Donald Rumsfeld, memo to Gen. Dick Myers et al., "Preparation of the Battlespace," *Rumsfeld Papers*, September 2, 2004, http://library.rumsfeld.com/doclib/sp/1314/To%20Gen.%20Dick%20Myers%20et%20al.%20re%20'Preparation%20of%20the%20Battlespace'%2009-02-2004.pdf#search="Preparation of the Battlespace.
18. "Advance Force Operations Contract," *The Intercept*, October 15, 2015, https://theintercept.com/document/2015/10/15/advance-force-operations-contract/.
19. IBM, National Interest Security Corporation, "An Introduction to

Edge Methods," June 17, 2010, https://www.documentcloud.org/documents/2483499-edge-methods-ibm-bao-intell-v4.html.

20. Adam Entous, "Obama Kept Looser Rules for Drones in Pakistan," *Wall Street Journal*, April 26, 2015, http://www.wsj.com/articles/obama-kept-looser-rules-for-drones-in-pakistan-1430092626.
21. Ali Watkins, "Obama Administration on Plan to Take Away CIA's Drones: Never Mind, Keep 'Em," *Huffington Post*, June 24, 2015, http://www.huffingtonpost.com/2015/06/24/obama-cia-drones_n_7649702.html.
22. Gordon Lubold, "Pentagon to Sharply Expand U.S. Drone Flights over Next Four Years," *Wall Street Journal*, August 16, 2015, http://www.wsj.com/articles/pentagon-to-add-drone-flights-1439768451.

THE KILL CHAIN

1. Jo Becker and Scott Shane, "Secret 'Kill List' Tests Obama's Principles," *New York Times*, May 29, 2012, http://www.nytimes.com/2012/05/29/world/obamas-leadership-in-war-on-al-qaeda.html?_r=0&mtrref=undefined&gwh=2D8CC1A187B950A05E9020C53E02197E&gwt=pay.
2. Chris Woods, "Obama's Five Rules for Covert Drone Strikes," Bureau of Investigative Journalism, September 6, 2012, https://www.thebureauinvestigates.com/2012/09/06/obamas-five-rules-for-covert-drone-strikes/.
3. Kimberly Dozier, "Who Will Drones Target? Who in the US Will Decide?," Big Story, May 22, 2012, http://bigstory.ap.org/content/who-will-drones-target-who-us-will-decide; Greg Miller, "Plan for Hunting Terrorists Signals U.S. Intends to Keep Adding Names to Kill Lists," *Washington Post*, October 23, 2012, https://www.washingtonpost.com/world/national-security/plan-for-hunting-terrorists-signals-us-intends-to-keep-adding-names-to-kill-lists/2012/10/23/4789b2ae-18b3-11e2-a55c-39408fbe6a4b_story.html.
4. Matthew Cole, Richard Esposito, and Jim Miklaszewski, "Why the White House Blessed the Recent Yemen Drone Strikes," *NBC News*, August 16, 2013, http://www.nbcnews.com/news/other/why-white-house-blessed-recent-yemen-drone-strikes-f6C10936036.
5. T. Mark McCurley and Kevin Maurer, *Hunter Killer: Inside America's Unmanned Air War* (New York: Dutton, 2015).
6. Scott Shane, "Yemen's Leader, President Hadi, Praises U.S. Drone

Strikes," *New York Times*, September 29, 2012, http://www.nytimes.com/2012/09/29/world/middleeast/yemens-leader-president-hadi-praises-us-drone-strikes.html?mtrref=undefined&gwh=F2AF3FF3DE686EFA14885DF28891156F&gwt=pay.

7. Zaid al-Alayaa and Paul Richter, "U.S. Resumes Drone Strikes in Yemen as Houthis Tighten Control," *Los Angeles Times*, February 6, 2015, http://www.latimes.com/world/middleeast/la-fg-yemen-houthis-20150206-story.html.
8. Congressional Research Service, "Memorandum: Legal Issues Related to the Lethal Targeting of U.S. Citizens Suspected of Terrorist Activities," May 4, 2012, https://www.propublica.org/documents/item/425630-crs-targeted-killing#document/p12/a71709.
9. White House, Office of the Press Secretary, "Fact Sheet: U.S. Policy Standards and Procedures for the Use of Force in Counterterrorism Operations outside the United States and Areas of Active Hostilities," May 23, 2013, https://www.whitehouse.gov/the-press-office/2013/05/23/fact-sheet-us-policy-standards-and-procedures-use-force-counterterrorism.
10. White House, Office of the Press Secretary, "Remarks by the President at the National Defense University," May 23, 2013, https://www.whitehouse.gov/the-press-office/2013/05/23/remarks-president-national-defense-university.
11. Kevin Jon Heller, "No, the UN Has Not Said the U.S. Is Engaged in an 'Armed Conflict' with Al Qaeda," *Opinio Juris*, May 21, 2011, http://opiniojuris.org/2011/05/21/no-the-un-has-not-affirmed-that-the-us-is-engaged-in-an-armed-conflict-with-al-qaeda/; Human Rights Watch, "US: Reassess War Model against Al Qaeda," July 31, 2013, https://www.hrw.org/news/2013/07/31/us-reassess-war-model-against-al-qaeda.
12. Bureau of Investigative Journalism, "US Strikes in Yemen, 2002 to Present," *undated*. https://docs.google.com/spreadsheets/d/1lb1hEYJ_omI8lSe33izwS2a2lbiygs0hTp2Al_Kz5KQ/edit?pli=1#gid=977256262.
13. Greg Miller and Karen DeYoung, "Brennan Nomination Exposes Criticism on Targeted Killings and Secret Saudi Base," *Washington Post*, February 5, 2013, https://www.washingtonpost.com/world/national-security/brennan-nomination-opens-obama-to-criticism-on-secret-targeted-killings/2013/02/05/8f3c94f0-6fb0-11e2-8b8d-e0b59a1b8e2a_story.html; Robert F. Worth,

Mark Mazzetti, and Scott Shane, "With Brennan Pick, a Light on Drone Strikes' Hazards," *New York Times*, February 5, 2013, http://www.nytimes.com/2013/02/06/world/middleeast/with-brennan-pick-a-light-on-drone-strikes-hazards.html?hp&pagewanted=all&_r=1&mtrref=undefined&gwh=0ADF0023E4C06F8649F2C135610E6D72&gwt=pay.

14. Mohammed Mukhashaf, "Yemen Air Strikes Foil Saudi Hostage Release – Negotiator," Reuters, April 23, 2012, http://uk.reuters.com/article/uk-saudi-yemen-kidnappers-idUKBRE83M15020120423.
15. Greg Miller, "White House Approves Broader Yemen Drone Campaign," *Washington Post*, April 25, 2012, https://www.washingtonpost.com/world/national-security white-house-approves-broader-yemen-drone-campaign/2012/04/25/gIQA82U6hT_story.html; Daniel Klaidman, "Drones: The Silent Killers," *Newsweek*, May 28, 2012, http://www.newsweek.com/drones-silent-killers-64909.
16. Gregory S. McNeal, "Targeted Killing and Accountability," *Georgetown Law Journal* 102 (2014): 681.
17. Karen DeYoung, "CIA Veteran John Brennan Has Transformed U.S. Counterterrorism Policy," *Washington Post*, October 24, 2012, https://www.washingtonpost.com/world/national-security/cia-veteran-john-brennan-has-transformed-us-counterterrorism-policy/2012/10/24/318b8eec-1c7c-11e2-ad90-ba5920e56eb3_story.html.
18. Human Rights Watch, "'Between a Drone and Al-Qaeda': The Civilian Cost of US Targeted Killings in Yemen," October 22, 2013, https://www.hrw.org/report/2013/10/22/between-drone-and-al-qaeda/civilian-cost-us-targeted-killings-yemen; Amrit Singh, "Death by Drone: Civilian Harm Caused by U.S. Targeted Killings in Yemen," Open Society Justice Initiative, 2015, https://www.opensocietyfoundations.org/sites/default/files/death-drones-report-eng-20150413.pdf.
19. Jeremy Diamond, "U.S. Drone Strike Accidentally Killed 2 Hostages," *CNN*, April 23, 2015, http://www.cnn.com/2015/04/23/politics/white-house-hostages-killed/.

THE HEART OF THE DRONE MAZE

1. German Bundestag, "Act to Introduce the Code of Crimes against International Law," June 26, 2002, http://www.iuscomp.org/gla/statutes/VoeStGB.pdf.

2. "Bündnisse: Der Krieg via Ramstein," *Der Spiegel*, April 17, 2015, http://www.spiegel.de/politik/deutschland/ramstein-air-base-us-drohneneinsaetze-aus-deutschland-gesteuert-a-1029264.html.
3. Alice K. Ross, "Who Is Dying in Afghanistan's 1,000-Plus Drone Strikes?," Bureau of Investigative Journalism, July 24, 2015, https://www.thebureauinvestigates.com/2014/07/24/who-is-dying-in-afghanistans-1000-plus-drone-strikes/; "Get the Data: Drone Wars," Bureau of Investigative Journalism, https://www.thebureauinvestigates.com/category/projects/drones/drones-graphs/.
4. "US-Drohnenkrieg läuft über Deutschland," *ARD*, May 30, 2013, http://daserste.ndr.de/panorama/archiv/2013/ramstein109.html%20; Christian Fuchs, John Goetz, and Hans Leyendecker, "Exclusive: US Armed Forces Piloting Drones from Bases In Germany," *Süddeutsche Zeitung*, June 6, 2013, http://international.sueddeutsche.de/post/52323491304/exclusive-us-armed-forces-piloting-drones-from.
5. German Bundestag, "On the Role of The United States Africa Command Stationed in Germany in Targeted Killings by US Armed Forces in Africa," Bundestag Printed Paper 17/14047, June 14, 2013 (question), and Bundestag Printed Paper 17/14401, July 18, 2013 (answer), http://goo.gl/ERbp9f.
6. "Death by Drone: Civilian Harm Caused by U.S. Targeted Killings in Yemen," Open Society Foundations, April 2015, https://www.opensocietyfoundations.org/reports/death-drone.
7. Reprieve, "Drone Victims Sue German Government for Facilitating Strikes in Yemen," October 15, 2014, http://www.reprieve.org.uk/press/2014_10_15_drone_victims_sue_german_govt/.
8. Intelsat, "Intelsat Repositions Satellite to Serve Military Units in Asia and Mideast," March 24, 2009, http://www.intelsat.com/news/intelsat-repositions-satellite-to-serve-military-units-in-asia-mideast/.
9. "Report on the Technical Expansion of the European Technical Center in the Mainz-Kastel Neighborhood of Wiesbaden," *Der Spiegel*, June 18, 2014, http://www.spiegel.de/media/media-34083.pdf.

TARGET AFRICA

1. John Vandiver, "Staging Sites Enable AFRICOM to Reach Hot Spots 'within 4 Hours,' Leader Says," *Stars and Stripes*, May 8, 2015, http://www.stripes.com/news/africa/staging-sites-enable-africom-to-reach-hot-spots-within-4-hours-leader-says-1.345120.
2. Luis Ramirez, "US Military Pays Close Attention to Boko Haram Militants," Voice of America, June 13, 2013, http://www.voanews.com/content/us-military-pays-close-attention-to-boko-haram-militants/1681488.html; Donna Miles, "Mullen Visits Horn of Africa Task Force," U.S. Department of Defense, February 24, 2011, http://archive.defense.gov/news/newsarticle.aspx?id=62927.
3. Luis Ramirez, "US Military Relocates Fleet from Djibouti Base," Voice of America, September 25, 2013, http://www.voanews.com/content/us-moves-drones-from-key-africa-base/1756946.html; Somturk, "Djibouti Craft Apron Parking Area Construction Logistics," http://somturk.com.tr/djibouti-craft-apron-parking-area-construction-logistics; Agence France-Presse, "US Relocates Drones Airfield after Djibouti Crashes," *Defense News*, September 25, 2013, http://archive.defensenews.com/article/20130925/DEFREG02/309250035/US-Relocates-Drones-Airfield-After-Djibouti-Crashes.
4. Craig Whitlock, "Pentagon Set to Open Second Drone Base in Niger As It Expands Operations in Africa," *Washington Post*, September 1, 2014, https://www.washingtonpost.com/world/national-security/pentagon-set-to-open-second-drone-base-in-niger-as-it-expands-operations-in-africa/2014/08/31/365489c4-2eb8-11e4-994d-202962a9150c_story.html; Editorial Board, "Dealing with Boko Haram," *New York Times*, May 30, 2014, http://www.nytimes.com/2014/05/31/opinion/dealing-with-boko-haram.html?_r=0&mtrref=undefined&gwh=3B9A11449E35972F0C99E9D10DBABBFE&gwt=pay&assetType=opinion.
5. Craig Whitlock, "U.S. Drone Base in Ethiopia Is Operational," *Washington Post*, October 27, 2011, https://www.washingtonpost.com/world/national-security/us-drone-base-in-ethiopia-is-operational/2011/10/27/gIQAznKwMM_story.html.
6. Thom Shanker, "Simple ScanEagle Drones a Boost for U.S. Military," *New York Times*, January 24, 2013, http://www.nytimes.com/2013/01/25/us/simple-scaneagle-drones-a-boost-for

-us-military.html?_r=0&mtrref=undefined&gwh=692024C7AFF317A04E929AFE55BF5CA9&gwt=pay; Naval Air Systems Command, "MQ-8 Fire Scout," http://www.navair.navy.mil/index.cfm?fuseaction=home.display&key=8250AFBA-DF2B-4999-9EF3-0B0E46144D03.

7. Ty McCormick, "Exclusive: U.S. Operates Drones from Secret Bases in Somalia," *Foreign Policy*, July 2, 2015, http://foreignpolicy.com/2015/07/02/exclusive-u-s-operates-drones-from-secret-bases-in-somalia-special-operations-jsoc-black-hawk-down/.
8. Cryptome, "CIA-DoD Chabelly Djibouti Drone Base," May 23, 2015, https://cryptome.org/2014-info/chabelly/chabelly-drone-base.htm; James Kitfield, "U.S. Using Local Soldiers to Fight Al-Qaida Allies in East Africa," Nuclear Threat Initiative, March 11, 2013, http://www.nti.org/gsn/article/united-states-using-local-soldiers-fight-al-qaida-allies-east-africa/.
9. Nick Turse, "Tomgram: Nick Turse, AFRICOM Becomes a 'War-Fighting Combatant Command,'" *TomDispatch*, April 13, 2014, http://www.tomdispatch.com/blog/175830/tomgram%3A_nick_turse%2C_africom_becomes_a_%22war-fighting_combatant_command%22.
10. "Bill Summary & Status 114th Congress (2015–2016) H.R.1735," Library of Congress, http://thomas.loc.gov/cgi-bin/bdquery/z?d114:H.R.1735:; "Committee Reports 114th Congress (2015–2016) House Report 114-102," Library of Congress, http://thomas.loc.gov/cgi-bin/cpquery/?&sid=cp114llNjh&r_n=hr102.114&dbname=cp114&&sel=TOC_1188734&.
11. Craig Whitlock, "Pentagon Setting Up Drone Base in Africa to Track Boko Haram Fighters," *Washington Post*, October 14, 2015, https://www.washingtonpost.com/world/national-security/pentagon-setting-up-drone-base-in-africa-to-track-boko-haram-fighters/2015/10/14/0cbfac94-7299-11e5-8d93-0af317ed58c9_story.html; John Vandiver, "Staging Sites Enable AFRICOM to Reach Hot Spots 'within 4 Hours,' Leader Says," *Stars and Stripes*, May 8, 2015, http://www.stripes.com/news/africa/staging-sites-enable-africom-to-reach-hot-spots-within-4-hours-leader-says-1.345120.
12. Eliza Griswold, "Can General Linder's Special Operations Forces Stop the Next Terrorist Threat?," *New York Times Magazine*, June 13, 2014, http://www.nytimes.com/2014/06/15/magazine/can-general-linders-special-operations-forces-stop-the-next-terror

ist-threat.html?mtrref=undefined&gwh=CCB5AE7F55C992AF63D66DE0E5F5FC81&gwt=pay&assetType=nyt_now.

13. Joshua Horton, "Camp Lemonnier Goes Green, Installs Solar Panels to Lower Energy Costs," United States Africa Command, October 29, 2012, http://www.africom.mil/newsroom/article/10020/camp-lemonnier-goes-green-installs-solar-panels-to; US Africom Public Affairs, "U.S. Army National Guard Conducts Water Drilling Tests at Camp Lemonnier," United States Africa Command, March 21, 2012, http://www.africom.mil/newsroom/article/8871/us-army-national-guard-conducts-water-drilling-tes; Maria R. Escamilla, "Toby Keith Visits Camp Lemonnier," United States Africa Command, April 27, 2012, http://www.africom.mil/newsroom/article/8940/toby-keith-visits-camp-lemonnier.
14. U.S. Africom Public Affairs, "2012 Posture Statement: Statement of General Carter Ham before House Armed Services Committee," United States Africa Command, March 1, 2012, http://www.africom.mil/newsroom/article/8832/2012-posture-statement-statement-of-general-carter.
15. Hurlburt Field, "Four Hurlburt Airmen Die in U-28A Crash in Djibouti," February 20, 2012, http://www.hurlburt.af.mil/News/ArticleDisplay/tabid/136/Article/204913/four-hurlburt-airmen-die-in-u-28a-crash-in-djibouti.aspx; Lockheed Martin, "P-3 Orion," http://www.lockheedmartin.com/us/products/p3.html.
16. Craig Whitlock, "Remote U.S. Base at Core of Secret Operations," *Washington Post*, October 25, 2012, https://www.washingtonpost.com/world/national-security/remote-us-base-at-core-of-secret-operations/2012/10/25/a26a9392-197a-11e2-bd10-5ff056538b7c_story.html.
17. John Vandiver, "Military Responds to Air Safety Issues in Djibouti," *Stars and Stripes*, May 5, 2015, http://www.stripes.com/news/military-responds-to-air-safety-issues-in-djibouti-1.344317; Combined Joint Task Force–Horn of Africa, http://www.hoa.africom.mil/.
18. Jim Garamone, "Work Salutes U.S. Africa Command's Efforts," U.S. Department of Defense, April 17, 2015, http://www.defense.gov/News-Article-View/Article/604484; Steve Kriss, "Working with the Less Fortunate, Developing Medical and Leadership Skills in East Africa," *Navy Medicine Live*, June 2, 2015, http://navymedicine.navylive.dodlive.mil/archives/8884; Jim Garamone, "Building

a Presence in Djibouti," U.S. Department of Defense, December 11, 2002, http://archive.defense.gov/news/newsarticle.aspx?id=42401.

19. Nick Turse, "Tomgram: Nick Turse, AFRICOM's Gigantic 'Small Footprint,'" *TomDispatch*, September 5, 2013, http://www.tomdispatch.com/blog/175743/tomgram%3A_nick_turse,_africom's_gigantic_%22small_footprint%22.
20. Jeremy Scahill, "The Dangerous U.S. Game in Yemen," *The Nation*, March 30, 2011, http://www.thenation.com/article/dangerous-us-game-yemen/.
21. Josh Wood, "Djibouti, a Safe Harbour in the Troubled Horn of Africa," *The National*, June 2, 2015, http://www.thenational.ae/world/africa/djibouti-a-safe-harbour-in-the-troubled-horn-of-africa.

DEATH BY METADATA

1. Jack Serle, "More Than 2,400 Dead As Obama's Drone Campaign Marks Five Years," Bureau of Investigative Journalism, January 23, 2014, https://www.thebureauinvestigates.com/2014/01/23/more-than-2400-dead-as-obamas-drone-campaign-marks-five-years/.
2. Dana Priest, "NSA Growth Fueled by Need to Target Terrorists," *Washington Post*, July 21, 2013, https://www.washingtonpost.com/world/national-security/nsa-growth-fueled-by-need-to-target-terrorists/2013/07/21/24c93cf4-f0b1-11e2-bed3-b9b6fe264871_story.html.
3. Center for Constitutional Rights, "Al-Majalah Freedom of Information Act Request," April 17, 2012, http://ccrjustice.org/home/what-we-do/our-cases/al-majalah-freedom-information-act-request#.
4. Mollie Reilly, "Obama Told Aides He's 'Really Good at Killing People,' New Book 'Double Down' Claims," *Huffington Post*, November 3, 2013, http://www.huffingtonpost.com/2013/11/03/obama-drones-double-down_n_4208815.html.

FIRING BLIND

1. Greg Miller, "Obama's New Drone Policy Leaves Room for CIA Role," *Washington Post*, May 25, 2013, https://www.washingtonpost.com/world/national-security/obamas-new-drone-policy-has-cause-for-concern/2013/05/25/0daad8be-c480-11e2-914f-a7aba60512a7_story.html.

2. Michael T. Flynn, Rich Juergens, and Thomas L. Cantrell, "Employing ISR SOF Best Practices," Defense Technical Information Center, 2008, http://www.dtic.mil/doctrine/docnet/courses/intelligence/intel/jfq_50_art-2.pdf.
3. Craig Whitlock and Greg Miller, "U.S. Moves Drone Fleet from Camp Lemonnier to Ease Djibouti's Safety Concerns," *Washington Post*, September 24, 2013. https://www.washingtonpost.com/world/national-security/drone-safety-concerns-force-us-to-move-large-fleet-from-camp-lemonnier-in-djibouti/2013/09/24/955518c4-213c-11e3-a03d-abbedc3a047c_story.html.
4. Stanley McChrystal, *My Share of the Task* (New York: Penguin, 2013).
5. Michael T. Flynn, Rich Juergens, and Thomas L. Cantrell, "Employing ISR SOF Best Practices," Defense Technical Information Center, 2008, http://www.dtic.mil/doctrine/docnet/courses/intelligence/intel/jfq_50_art-2.pdf.
6. "Recommendations and Report of the Task Force on US Drone Policy," Stimson Center, 2014, http://www.stimson.org/images/uploads/task_force_report_final_web_062414.pdf.
7. Adam Entous, Julian E. Barnes, and Margaret Coker, "U.S. Doubts Intelligence That Led to Yemen Strike," *Wall Street Journal*, December 29, 2011, http://www.wsj.com/articles/SB10001424052970203899504577126883574284126.
8. Joint Chiefs of Staff, "Special Operations: Joint Publication 3-05," Defense Technical Information Center, July 16, 2014, http://www.dtic.mil/doctrine/new_pubs/jp3_05.pdf.
9. Aspen Institute, "A Look into SOCOM," *Aspen Security Forum*, n.d., http://aspensecurityforum.org/wp-content/uploads/2015/07/A-Look-into-SOCOM.pdf.

STINGRAYS AT HOME

1. "The Secret Surveillance Catalogue," *The Intercept*, December 17, 2015, https://theintercept.com/surveillance-catalogue/.
2. "Trott Leads Effort to Stand against Obama Administration's Weakening of Local Police Departments Ability to Respond to Terror Attacks," Dave Trott, House of Representatives, December 7, 2015, https://trott.house.gov/media-center/press-releases/trott-leads-effort-stand-against-obama-administration-s-weakening-local.
3. Kim Zetter, "Turns Out Police Stingray Spy Tools Can Indeed Record

Calls," *Wired*, October 28, 2015, http://www.wired.com/2015/10/stingray-government-spy-tools-can-record-calls-new-documents-confirm/.

4. Tsutomu Shimomura, "Catching Kevin," *Wired*, February 1, 1996, http://www.wired.com/1996/02/catching/.
5. Mark Hoover, "Top 100: Harris Digests Exelis Acquisition, Plans to Leverage New Scale," *Washington Technology*, June 11, 2015, https://washingtontechnology.com/articles/2015/06/07/harris-top-100-profile.aspx; "Entity Actions for 'Ventis Corporation,'" *Barbara K. Cegavske, Nevada Secretary of State*, http://nvsos.gov/sosentitysearch/corpActions.aspx?lx8nvq=kaC42PDAVuZhhwdFwAw4Fg%3d%3d&CorpName=VENTIS+CORPORATION.
6. Devlin Barrett, "CIA Aided Program to Spy on U.S. Cellphones," *Wall Street Journal*, March 10, 2015, http://www.wsj.com/articles/cia-gave-justice-department-secret-phone-scanning-technology-1426009924.
7. Federal Communications Commission, "Re: FOIA Control No. 2011-586," *Amazon News*, February 29, 2012, https://s3.amazonaws.com/files.cloudprivacy.net/FOIA/FCC/fcc-stingray-reply.pdf.
8. ACLU, "Stingray Tracking Devices: Who's Got Them?," https://www.aclu.org/map/stingray-tracking-devices-whos-got-them.
9. Justin Fenton, "Baltimore Police Used Secret Technology to Track Cellphones in Thousands of Cases," *Baltimore Sun*, April 9, 2015, http://www.baltimoresun.com/news/maryland/baltimore-city/bs-md-ci-stingray-case-20150408-story.html.
10. Nathan Freed Wessler, "Police Citing 'Terrorism' to Buy Stingrays Used Only for Ordinary Crimes," ACLU, October 23, 2015, https://www.aclu.org/blog/free-future/police-citing-terrorism-buy-stingrays-used-only-ordinary-crimes.
11. "Police Are Using a Powerful Surveillance Tool to Fight the War on Drugs, Not Terrorism," *PrivacySOS*, https://privacysos.org/blog/police-are-using-a-powerful-surveillance-tool-to-fight-the-war-on-drugs-not-terrorism/.
12. Florida Department of Law Enforcement, "Description of Intended Single Source Purchase," April 25, 2013, http://www.myflorida.com/apps/vbs/adoc/F13170_CopyofPUR77767dayIntendedSoleSourceSyndetix.pdf; ACLU, "Florida Department of Law Enforcement Stingray Purchase Order Summary," https://www.aclu.org/florida-department-law-enforcement-stingray-purchase-order

-summary?redirect=technology-and-liberty/florida-department-law-enforcement-stingray-purchase-order-summary.

13. Ali Winston, "East Bay Cellphone Surveillance Plan Gets Attorney General's Support," *Reveal News*, September 30, 2015, https://www.revealnews.org/article/east-bay-cellphone-surveillance-plan-gets-attorney-generals-support/.
14. U.S. Department of Justice and Federal Bureau of Investigation, "Re: Acquisition of Wireless Collection Equipment/Technology and Non-Disclosure Obligations," NYCLU, http://www.nyclu.org/files/20120629-renondisclosure-obligations(Harris-ECSO).pdf.
15. Department of Justice Office of Public Affairs, "Department Announces Enhanced Policy for Use of Cell-Site Simulators," September 3, 2015, http://www.justice.gov/opa/pr/justice-department-announces-enhanced-policy-use-cell-site-simulators.
16. U.S. Department of Homeland Security, "Memorandum: Department Policy Regarding the Use of Cell-Site Simulator Technology," October 19, 2015, https://oversight.house.gov/wp-content/uploads/2015/10/15-3959-S2-DHS-Signed-Policy-Directive-047-02-Use-of-Cell-Site-Simulator-Tech.pdf; Neema Singh Guliani, "The Four Biggest Problems with DHS's New Stingray Policy," ACLU, October 22, 2015, https://www.aclu.org/blog/free-future/four-biggest-problems-dhss-new-stingray-policy.
17. Iain D. Johnston, "Memorandum Opinion and Order: *United States of America v. In the Matter of the Application of the United States*," United States District Court Northern District of Illinois Western Division, November 9, 2015, https://www.documentcloud.org/documents/2516907-united-states-of-america-v-in-the-matter-of-the.html.

THE LIFE AND DEATH OF OBJECTIVE PECKHAM

1. Abdi Guled and Katharine Houreld, "Officials: US Drone Strike Killed Somali Insurgent," *San Diego Union-Tribune*, January 21, 2012, http://www.sandiegouniontribune.com/news/2012/jan/21/officials-us-drone-strike-killed-somali-insurgent/.
2. "The Shabab in Somalia: A Very British Execution?," *The Economist*, January 25, 2012, http://www.economist.com/blogs/baobab/2012/01/shabab-somalia.
3. WikiLeaks, "The Guantanamo Files: Abdul Malik Bajabu," April 24, 2011, https://wikileaks.org/gitmo/prisoner/10025.html.

4. Aaron Y. Zelin, "New Release: 'Biography of the Martyred Figures in East Africa #5: Bilāl al-Birjāwī al-Lubnānī (Abu ḤafṢ),'" *Jihadology*, April 15, 2013, http://jihadology.net/2013/04/15/new-release-biography-of-the-martyred-figures-in-east-africa-5-bilal-al-birjawi-al-lubnani-abu-%E1%B8%A5af%E1%B9%A3/.
5. UK Legislation, "Terrorism Act 200: Schedule 7," http://www.legislation.gov.uk/ukpga/2000/11/schedule/7.
6. Abdi Guled and Malkhdir M. Muhumed, "'Partner' Airstrike Hits Somali Militants' Convoy," *Huffington Post*, June 24, 2011, http://www.huffingtonpost.com/huff-wires/20110624/af-somalia/.
7. Malyun Ali, "Somalia: Nine Al-Shabab Members Killed in U.S. Air Raid in Southern Somalia," *Internet Archive*, July 6, 2011, https://web.archive.org/web/20110713055710/http://www.raxanreeb.com/?p=102606.
8. Abdalle Ahmed, "British Citizen Killed by US Military in Somalia," *Pan-African News Wire*, January 22, 2012, http://panafricannews.blogspot.fr/2012/01/british-citizen-killed-by-us-military.html.
9. Jason Burke and James Orr, "Al-Qaida Bomber Fazul Abdullah Mohammed Killed," *The Guardian*, June 11, 2011, http://www.theguardian.com/world/2011/jun/11/al-qaida-bomber-fazul-abdullah-mohammed-killed.
10. "Somalia: Al Qaeda Commander Killed in Blast," *Internet Archive*, January 24, 2012, https://web.archive.org/web/20121020083857/http://allafrica.com/stories/201201250052.html.
11. MHD, Rashid Nuune, "Al Queda, Al-Shaabab Pledge Allegiance . . . Again," *Somalia Report*, February 9, 2012, http://www.somaliareport.com/index.php/post/2749/Al_Qaeda_Al-Shaabab_Pledge_AllegianceAgain.
12. CNN Wire Staff, "Al-Shabaab Joining al Qaeda, Monitor Group Says," *CNN*, February 10, 2012, http://edition.cnn.com/2012/02/09/world/africa/somalia-shabaab-qaeda/.
13. Abdi Guled, "Al-Shabab Executes 3 Members," *Washington Times*, July 22, 2012, http://www.washingtontimes.com/news/2012/jul/22/al-shabab-executes-3-members/.
14. Chris Woods and Alice K. Ross, "Former British Citizens Killed by Drone Strikes after Passports Revoked," Bureau of Investigative Journalism, February 27, 2013, https://www.thebureauinvestigates.com/2013/02/27/former-british-citizens-killed-by-drone-strikes-after-passports-revoked/.

15. Victoria Parsons, "What Do We Know about Citizenship Stripping?" Bureau of Investigative Journalism, December 10, 2014, https://www.thebureauinvestigates.com/2014/12/10/what-do-we-know-about-citizenship-stripping/.
16. Jack Serle, "RAF Drone Strike: Syria Deaths Means at Least 10 Britons Now Killed by Drones in West's War on Terror," Bureau of Investigative Journalism, September 7, 2015, https://www.thebureauinvestigates.com/2015/09/07/raf-drone-strike-syria-deaths-means-at-least-10-britons-now-killed-by-drones-in-wests-war-on-terror/.
17. Kimiko De Freytas-Tamura, "Junaid Hussain, ISIS Recruiter, Reported Killed in Airstrike," *New York Times*, August 27, 2015, http://www.nytimes.com/2015/08/28/world/middleeast/junaid-hussain-islamic-state-recruiter-killed.html?_r=0&mtrref=undefined&gwh=20004848E1EC0B56D0FA8A15E789620B&gwt=pay.
18. "Ruhul Amin and Reyaad Khan: The Footballer and the Boy Who Wanted to Be First Asian PM," *The Guardian*, September 7, 2015, http://www.theguardian.com/world/2015/sep/07/british-isis-militants-killed-raf-drone-strike-syria-reyaad-khan-ruhul-amin.
19. Dominic Casciani, "Islamic State: Profile of Mohammed Emwazi aka 'Jihadi John,'" *BBC News*, November 13, 2015, http://www.bbc.com/news/uk-31641569.

MANHUNTING IN THE HINDU KUSH

1. Spencer Ackerman, "US Drone Strikes More Deadly to Afghan Civilians Than Manned Aircraft – Adviser," *The Guardian*, July 2, 2013, http://www.theguardian.com/world/2013/jul/02/us-drone-strikes-afghan-civilians.
2. "Operation Haymaker," *The Intercept*, October 15, 2015, https://theintercept.com/document/2015/10/14/operation-haymaker/#page-16.
3. "ISAF Joint Command Morning Operational Update, Nov 9," *RS News*, November 9, 2012, http://www.rs.nato.int/article/isaf-releases/isaf-joint-command-morning-operational-update-nov-9.html.
4. Elizabeth Rubin, "Battle Company Is Out There," *New York Times Magazine*, February 24, 2008, http://www.nytimes.com/2008/02/24/magazine/24afghanistan-t.html.

5. Matt Trevithick and Daniel Seckman, "Heart of Darkness: Into Afghanistan's Taliban Valley," *Daily Beast*, November 15, 2014, http://www.thedailybeast.com/articles/2014/11/15/heart-of-darkness-into-afghanistan-s-korengal.html.
6. "SEAL Team 6: A Secret History of Quiet Killings and Blurred Lines," *New York Times*, June 6, 2015, http://www.nytimes.com/2015/06/07/world/asia/the-secret-history-of-seal-team-6.html?mtrref=undefined&gwh=EFFAF7F2997B98F35D5C5F3855636E98&gwt=pay.
7. Gary Emery, "New Hurlburt Intel Squadron Turns Aerial Eye on Terrorists," Air Force Special Operations Command, August 11, 2006, http://www.afsoc.af.mil/News/ArticleDisplay/tabid/136/Article/163478/new-hurlburt-intel-squadron-turns-aerial-eye-on-terrorists.aspx.
8. United Nations Assistance Mission In Afghanistan, http://unama.unmissions.org/.
9. International Human Rights and Conflict Resolution Clinic (Stanford Law School) and Global Justice Clinic (NYU School of Law), "Living Under Drones: Death, Injury, and Trauma to Civilians from Drone Practices in Pakistan," Center for Human Right and Global Justice, NYU School of Law, September 2012, http://chrgj.org/wp-content/uploads/2012/10/Living-Under-Drones.pdf.
10. Alissa J. Rubin, "Karzai to Forbid His Forces to Request Foreign Airstrikes," *New York Times*, February 16, 2013, http://www.nytimes.com/2013/02/17/world/asia/karzai-to-forbid-his-forces-from-requesting-foreign-airstrikes.html?mtrref=undefined&gwh=95621A8600168B50C65B22BDDBC48580&gwt=pay.
11. Cora Currier, "Our Condolences: How the U.S. Paid for Death and Damage in Afghanistan," *The Intercept*, February 27, 2015, https://theintercept.com/2015/02/27/payments-civilians-afghanistan/.
12. Doctors Without Borders, "'Unspeakable': An MSF Nurse Recounts the Attack on MSF's Kunduz Hospital," October 3, 2015, http://www.doctorswithoutborders.org/article/unspeakable-msf-nurse-recounts-attack-msfs-kunduz-hospital.
13. Micah Zenko, "Kunduz Airstrike and Civilian Deaths in Afghanistan," Council on Foreign Relations, October 3, 2015, http://blogs.cfr.org/zenko/2015/10/03/kunduz-airstrike-and-civilian-deaths-in-afghanistan/; Defense Casualty Analysis System, "U.S. Military Casualties – Operation Enduring Freedom (OEF) Casualty Summary by Casualty Category," DMDC, December 24, 2015,

https://www.dmdc.osd.mil/dcas/pages/report_oef_type.xhtml; Neta C. Crawford, "Afghan Civilians," Watson Institute, March 2015, http://watson.brown.edu/costsofwar/costs/human/civilians/afghan.

14. UN News Centre, "Civilians Continue to Bear Brunt of Afghan Conflict, New UN Report Reveals," August 5, 2015, http://www.un.org/apps/news/story.asp?NewsID=51563#.VnxWFZMrLBK.
15. Sean D. Naylor, "Insurgent Bombings Rise as U.S. Eases off the Taliban," *Foreign Policy*, August 14, 2015, http://foreignpolicy.com/2015/08/14/easing-off-the-taliban/.
16. Wesley Morgan, "Special Report: Ten Years in Afghanistan's Pech Valley," United States Institute of Peace, September 2015, http://www.usip.org/sites/default/files/SR382-Ten-Years-in-Afghanistans-Pech-Valley.pdf.
17. Greg Miller, "U.S. Launches Secret Drone Campaign to Hunt Islamic State Leaders in Syria," *Washington Post*, September 1, 2015, https://www.washingtonpost.com/world/national-security/us-launches-secret-drone-campaign-to-hunt-islamic-state-leaders-in-syria/2015/09/01/723b3e04-5033-11e5-933e-7d06c647a395_story.html.

GLOSSARY

A

ABI
activity-based intelligence

ADNI/SRA
assistant director of national intelligence for systems and resource analyses

AF/AFG
Afghanistan

AFO
advance force operation

AFRICOM
U.S. Africa Command

ANSF
Afghan National Security Forces

AO
area of operations

AOR
area of responsibility

AP
Arabian Peninsula

APG
aerial precision geolocation

AQ
al Qaeda

AQ FAC
al Qaeda facilitator

AQIM
al Qaeda in the Islamic Maghreb

AQSL
al Qaeda senior leadership

AUMF
Authorization for Use of Military Force (2001)

B

BBC
baseball card

BDA
battle damage assessment

BDL
bed-down location

C

CAOC
Combined Air and Space Operations Center

CAP
Combat Air Patrol

CAPE
Department of Defense Office of Cost Assessment and Program Evaluation

CCMD
Combatant Command

CDE
collateral damage environment

CELLEX
cellular exploitation

CF
Coalition Forces

CIVCAS
civilian casualties

CJSOTF
Combined Joint Special Operations Task Force

C-LRA
Counter-Lord's Resistance Army

CNO
computer network operations

CoM
chief of mission (ambassador)
COMINT
communications intelligence
CONOP
concept of operations
CoS
chief of station (CIA)
CT
counterterrorism

D
DNR COMINT
dial number recognition communications intelligence
DOD
Department of Defense
DOMEX
document and media exploitation

E
EA
East Africa
EKIA
enemy killed in action
EWIA
enemy wounded in action

F
F3EA
find, fix, finish, exploit/analyze
FFF
find, fix, finish
FMV or **HD FMV**
full-motion video or high-definition full-motion video
FOB
Forward Operating Base
FS
Fire Scout (unmanned helicopter)
FVEY
Five Eyes

G
GCC
Geographic Combatant Command
GOCO
government-owned, contractor-operated
GSM
global system for mobile communication

H
HN
host nation
HOA
Horn of Africa
HUJI
Harkat-ul-Jihad-al-Islami
HUMINT or **HI**
human intelligence
HVI
high-value individual

I
IC
intelligence community
IED
improvised explosive device
IIR
intelligence information report

IMINT
imagery intelligence
ISAF
International Security Assistance Force
IPB
intelligence preparation of the battle space
IPOE
intelligence preparation of the operational environment
ISR
intelligence, surveillance, and reconnaissance
IZ
Iraq

J
J2
Intelligence Directorate of a Joint Staff
J5
Plans Directorate of a Joint Staff
JP
jackpot
JPEL
joint prioritized effects list
JTL
joint target list
JWICS
Joint Worldwide Intelligence Communications System

L
LeT
Lashkar-e-Taiba

M
MAM
military-age male
MARSOC
Marine Corps Special Operations Command
MC-12
manned surveillance aircraft
MEDEX
media exploitation
MFW
medium fixed-wing (manned aircraft)
MQ-1
General Atomics Predator drone
MQ-9
General Atomics Reaper drone

N
NAI
named area of interest
NAVAF
U.S. Naval Forces Africa
NSB
Yemeni National Security Bureau

O
OBJ
objective
ODTAAC
outside a defined theater of active armed conflict
OEF
Operation Enduring Freedom
OIF
Operation Iraqi Freedom

OPSEC
operations security
OPTEMPO
operational tempo
OSD
Office of Secretary of Defense

P
P-3
P-3 Orion (manned surveillance plane)
PACOM
U.S. Pacific Command
PAX
passengers
PDC/PC
Principals' Deputies Committee / Principals Committee
PID
positive identification
PKM
machine gun
POL
pattern of life
POTUS
president of the United States
PTT COMINT
push-to-talk communications intelligence

R
ROE
rules of engagement
ROYG
Republic of Yemen Government
RPA
remotely piloted aircraft (drone)

S
SE
ScanEagle (surveillance drone)
SECDEF
secretary of defense
SIGINT or **SI**
signals intelligence
SITREP
situation report
SME
subject matter expert
SNA
social network analysis
SOCOM
U.S. Special Operations Command
SRA
See **ADNI/SRA**

T
TADS
terror attack disruption strikes
TB
Taliban
TF
task force
TIR
tactical interrogation report
TTP
tactics, techniques, procedures; Tehrik-e Taliban Pakistan

U
U-28
manned surveillance plane

V

VBIED
vehicle-borne improvised explosive device

VEO
violent extremist organization

VID
voice identification

CONTRIBUTORS

Jeremy Scahill is one of the three founding editors of *The Intercept*. He is the author of *Dirty Wars: The World Is a Battlefield* and *Blackwater: The Rise of the World's Most Powerful Mercenary Army*. He was twice awarded the George Polk Award, in 1998 for foreign reporting and in 2008 for *Blackwater*. Scahill is a producer and writer of the film *Dirty Wars*, which premiered at the 2013 Sundance Film Festival and was nominated for an Academy Award.

Glenn Greenwald is one of the three founding editors of *The Intercept*. He is the author of four *New York Times* bestselling books on politics and law. His most recent book, *No Place to Hide*, is about the U.S. surveillance state and his experiences reporting on the Snowden documents around the world. The NSA reporting he led for *The Guardian* was awarded the 2014 Pulitzer Prize for public service.

Josh Begley is a research editor and data artist at *The Intercept*. He is the creator of Metadata+, an iPhone app that tracks U.S. drone strikes.

Cora Currier is a staff reporter at *The Intercept* with a focus on national security, foreign affairs, and human rights.

Ryan Devereaux is a staff reporter at *The Intercept*. His work has focused on organized crime, conflict, and human rights in the context of counterterrorism and the drug war.

Ryan Gallagher, a staff reporter at *The Intercept*, is an award-winning Scottish journalist whose work focuses on government surveillance, technology, and civil liberties.

Peter Maass, a senior editor at *The Intercept*, is the author of *Love Thy Neighbor: A Story of War*, an award-winning memoir about the conflict in Bosnia, and *Crude World: The Violent Twilight of Oil*.

Nick Turse is a frequent contributor to *The Intercept*. He is the managing editor of TomDispatch.com and the author, most recently, of *Tomorrow's Battlefield: U.S. Proxy Wars and Secret Ops in Africa* and *The*

Changing Face of Empire: Special Ops, Drones, Spies, Proxy Fighters, Secret Bases, and Cyberwarfare.

Margot Williams is the research editor for investigations at *The Intercept.* During fourteen years at the *Washington Post* she was a member of two Pulitzer Prize–winning teams, for a 1998 investigation of Washington, D.C., police shootings of civilians and in 2001 for national coverage of terrorism. Williams compiled the first list of the Guantánamo detainees – years before their names were made public – and created the comprehensive Guantánamo database for the *New York Times.*

ABOUT THE INTERCEPT AND FIRST LOOK MEDIA

The Intercept, a digital magazine founded by Glenn Greenwald, Laura Poitras, and Jeremy Scahill, is dedicated to producing fearless, adversarial journalism. We believe journalism should bring transparency and accountability to powerful institutions, and our journalists have the editorial freedom and legal support to pursue this mission.

The Intercept is a publication of First Look Media. Launched in 2013 by eBay founder and philanthropist Pierre Omidyar, First Look Media is a multiplatform media company devoted to supporting independent voices, from investigative journalism and documentary filmmaking to arts, culture, media, and entertainment. First Look Media produces and distributes content in a wide range of forms, including feature films, short-form video, podcasts, interactive media, and long-form journalism, for its own digital properties and with partners.

IMAGE CREDITS

p. 1: Saul Loeb/AFP/Getty Images.
p. 3: *The Intercept.*
p. 7: U.S. Air Force.
pp.14–15: Pablo Martinez Monsivai/AP.
pp.40–41: Veronique de Viguerie/Getty Images.
p.46: Haraz N. Ghanbari/AP.
p.47: Win McNamee/Getty Images.
p.48: Awlaki family.
p.49: Michael Reynolds/EPA/Landov.
p.58: Yuri Kadobnov/AFP/Getty Images, Scott Olson/Getty Images, Paul J. Richards/AFP/Getty Images, Saul Loeb/Pool/Getty Images, Mark Sagliocco/FilmMagic/Getty Images, Taylor Hill/Getty Images, Mark Wilson/Getty Images, Bryan Steffy/Getty Images, Paul Morigi/Getty Images.
p.59: Andrew Harrer/Bloomberg via Getty Images, Saul Loeb/Pool/Getty Images, Chip Somodevilla/Getty Images, Mark Wilson/Getty Images, Paul Morigi/Getty Images, Mark Wilson/Getty Images, Brendan McDermid/Reuters/Landov, Sean Gallup/Getty Images.
pp.68–69: Veronique de Viguerie/Getty Images.
p.74: Chris Hyde/Getty Images.
p.78: AFP/Getty Images.
pp.84–85: Map data: Google, DigitalGlobe.
pp.94–95: Kirsty Wigglesworth/AP.
p.117: Sami al-Ansi/AFP/Getty Images.
pp.134–135, p.138, p.139, p.145, p.148, p.150: Andrew Testa for *The Intercept.*
pp.152–153: Ed Darack/Getty Images.
p.146: U.S. Navy.
p.160: U.S. Department of Defense.
p.165: U.S. Army Cyber Command.
p.174: Namatullah Karyab/AFP/Getty Images.

INDEX

Page numbers in *italics* refer to illustrations.